EVERYTHING THAT CREEPETH UPON THE EARTH

"... any man who hath seen any
or the least of these
hath seen God
moving in his
majesty and power."

Doctrine and Covenants 88:47

by

Gary McCallister

EVERYTHING THAT CREEPETH ON THE EARTH

Copyright © 2023 Gary McCallister
All rights reserved.
ISBN:

DEDICATION

This book is for you. Yes, you, the reader. Step outside. Draw a circle around you with a ten-foot radius. Inside this circle are four hundred things no one understands. In fact, some, no one has ever even seen. Can you understand them? Can you find them?

CONTENTS

Acknowledgments	i
Preface	7
Part 1 Creeping things overview	21
Chapter 1 - Introduction	21
Chapter 2 - What really bugs me	29
Chapter 3 - A rose by any other name	39
Chapter 4 - Significance of creeping things	37
Chapter 5 - Neglected scientists	52
Part 2 - The Creepers	56
Chapter 6 - Blattidae (cockroaches)	58
Chapter 7 - Coleoptera (beetles)	69
Chapter 8 - Hymenoptera (bees)	90
Chapter 9 - Odonata	111
Chapter 10 - Orthoptera	118
Chapter 11 - Lepidoptera	132
Chapter 12 - Diptera	142
Chapter 13 - Mantidae	156
Chapter 14 - Isopoda	169
Chapter 15 - Conclusions	174

EVERYTHING THAT CREEPETH ON THE EARTH

ACKNOWLEDGMENTS

For all those people that introduced me to the world of the tiny creeping things. There are probably too many to list, but there are those who stand out in my memory. Perhaps the first in my memory is Dr. Lee F. Braithwaite, a professor of Zoology at Brigham Young University. He taught a course in Invertebrate Zoology the first year I was at that University. He was followed by Dr. Ferron Andersen also at BYU who introduced me to my lifelong interest in parasites. Dr. Vernon J Tipton taught Entomology, both general and medical. Dr, Gerald Schmidt at the University of Northern Colorado rounded out my education in the creeping things of the world. I have been engrossed in a fascinating and invisible world for over fifty years and I am grateful for the interesting life it has led me to.

PREFACE

"Teach ye diligently . . . Of things both in heaven and in the earth, and under the earth; . . . "
Doctrine and Covenants 88:78-80

The purpose of this book is to:
1. Testify of God's majesty and power.
2. Increase interest in the fundamental, and often forgotten, foundation of our earth, the little creatures.
3. Provide useful and interesting knowledge to people concerning a fundamental and interesting part of their world.
4. Encourage scientific interest and activities in people.

This preface explains further the purposes and background for this collection of writings about creeping things. If you are looking for excuses and justification for your morbid fascinations with creeping things, these ideas may be useful.

What we don't know

It is understood that books or any other media can only tell you the things we know. But the truth about the earth, the universe, and God, is that we know almost nothing. That's why I feel adequate writing this book. I know almost nothing and hope to learn more.

The good news is that I hope I can encourage people to learn more on their own. In fact, compared to the things that would be required to know to create even a successful terrarium, I know almost nothing. The bad news, those are my credentials. But the good news is, those are everyone's credentials and there is a magnificent world to learn about, understand, and care for, as humans have been given dominance over this marvelous creation.

Humanity is already fascinated by the higher animals. We name sports teams after predators and carnivores even though most of them are gangsters and bullies. That is, they hunt in packs of pick on things smaller than themselves. Humans are also narcissists with an entire field of study devoted to their own health and behavior. Plants have their own advocates and the beauty of flowers and greenery in gardens is appreciated by many. But there is a world in between upon which all the rest relies: the hidden world of the "creeping things".

There is a universe between the plants and the vertebrate animals. It is filled with tiny things like bacteria, protozoa, fungus, nematodes, earthworms, insects, and a handful of other similar, cryptic creatures. And <u>all that is obvious about the world relies on these creatures doing their jobs.</u>

These are the creatures that decompose organic back into soil. These are they that break stone into soil. These are they that recycle nutrients. These are they that control populations of invaders. These are they that pollinate plants and provide food for other organisms. It's the invisible, below the ground foundation of our world.

Huge numbers are involved and an overwhelming diversity of forms and lifestyles. Compared to the other life forms, we know relatively little about the individuals involved. Why do earthworms never get skin infections when they live in an environment surrounded by organisms that can infect human skin? What do mosquitoes eat when they are not requiring a blood meal? How many nematodes are there per gram in various soil types? The diagram below shows only a portion of the interactions that take place invisible to you and me.

There is a world of fascination awaiting exploration. What is it waiting for? You! The big boys and girls of science live in a world of public or corporate funding. New, powerful toys inspire them on to new discoveries before they even understand the world with the old toys. And the old toys are at your fingertips for a little effort, time, and very few dollars. Private science is cheaper by far than private entertainment.

It also doesn't need to be as lonely as you may think. Getting together with others to share your discoveries, equipment, ideas, and sociability is a distinct possibility. Never heard of such a thing. Start a science club yourself.

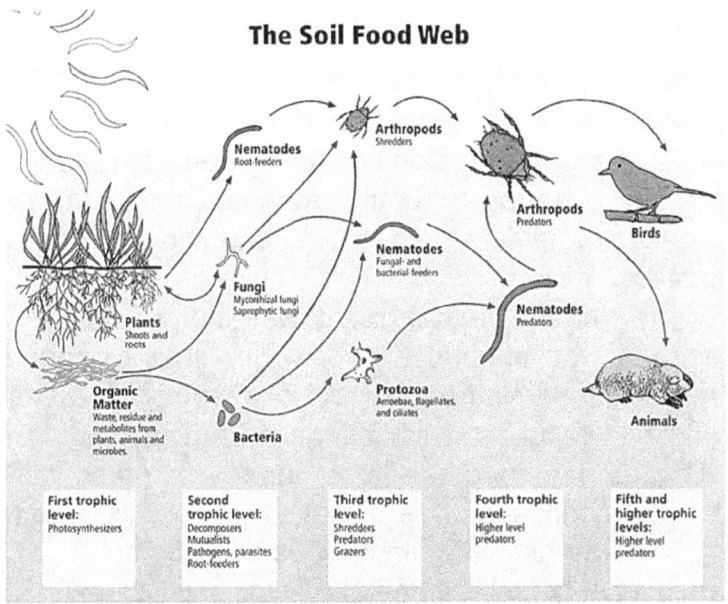

Relationships between soil food web, plants, organic matter, and birds and mammals
Image courtesy of USDA Natural Resources Conservation Service
http://soils.usda.gov/sqi/soil_quality/soil_biology/soil_food_web.html.

I have written a previous book on Nematodes with some of the same ideas and goals that this book tries to do with arthropods. See <u>Nematode Science Project Book:</u> How to find and study nematodes in the soil. 2021. Amazon.

"And Solomon's wisdom excelled the wisdom of all the children of the east country, and all the wisdom of Egypt. For he was wiser than all men; than Ethan the Ezrahite, and Heman, and Chalcol, and Darda, the sons of Mahol: and his fame was in all nations round about. And he spake three thousand proverbs: and his songs were a thousand

and five. And he spake of trees, from the cedar tree that is in Lebanon even unto the hyssop that springeth out of the wall: he spake also of beasts, and of fowl, **and of creeping things**, *and of fishes. And there came of all people to hear the wisdom of Solomon, from all kings of the earth, which had heard of his wisdom."*
1 Kings 4:30-34

<u>Who am I to tell you what to do?</u>

Well, I am not Solomon, but I did go to college. However, that is not why I should tell you what to do. In fact, that may be why I shouldn't be telling you what to do. College turns out to be more about how smart men are, and have been, and very little about God and the universe he has created.

But what I started to say was that, because I went to college, I was imbued with the idea that I needed to do something for which people will pay money. They told me it would be useful for such things as eating, finding shelter, getting married and raising a family. In other words, the purpose of education was to make me "socially useful".

Unfortunately, college doesn't really do that. Oh, you learn a lot of stuff, but none of it helps you DO anything. They say they are teaching you to think. I think thinking may not be what they think it is. I think plumbers must think clearly to fix your plumbing. Electricians must think clearly. If electricians don't think clearly, it's shocking, perhaps deadly. The people who seem to have the most trouble thinking clearly are the college professors who are trying to help the rest of us think clearly.

Anyway, I learned, through observation, not college, that the characteristic reaction of most people to creeping things is to eradicate them. I then discovered that people are generally willing to pay for this. However, learning about eradication leads one down the path to become, at least, a serial killer, and eventually a mass murderer. Perhaps not of humans, or at least not directly, but if you kill enough stuff it's bound to influence humans eventually.

I succumbed to this practice for many years and have caused the deaths of an unknown, but exceptionally large number of mosquitoes, and the odd cricket that my wife hates. By doing this I managed to keep my children alive.

I ran a mosquito control district for seventeen years while simultaneously teaching at the local university where we sometimes killed insects and the occasional rat. As a result, there is a price on my head by the insect mafia, and I am now the selected target for every mosquito known to man. Not to mention the PTSD, horrible dreams, and guilt for the things I have done. The eye twitch alone is miserable.

I tried to balance my militaristic actions by nurturing students for many years as a professor of biology. Unfortunately, it was sometimes difficult to tell if: 1) students profited from my lectures and exams, 2) were tortured by my instruction, or 3) were simply indoctrinated into continuing the deadly spiral. But at least I helped them think clearly.

Haunted by my violent past, I have tried to redeem myself by becoming a beekeeper. The theory is that if I raise as many generations of insects as I killed, the karma will balance out in the end. My wife thinks my addiction to peanut butter and honey sandwiches provides an alternative motive that disqualifies me partially from a karma balance.

I also spent several years studying a nearly invisible group of organisms called nematodes; unsegmented round worms that enrich the soil, but also sometimes parasitize animals, and damage crops. That was a thrill few people get to experience, let me tell you. I discovered a novel way of killing unborn nematodes. I can't decide if I am proud of inventing a method of nematode abortion, or ashamed.

But I digress. Nematodes are a topic of another book I have written. (Nematode Science Project Book for kids, teachers, and amateur scientists: How to find and study nematodes in the soil. Amazon.)

This tendency to eliminate or eradicate creeping things is deemed a meritorious action by many. It may be an innate human tendency. I should note that ants also pull the legs and wings off any insect that falls into their clutches, so it is not necessarily a unique human failing.

I also assure you that despite my discoveries, nematodes are in no danger of being eradicated. One cow pile can yield literally millions of baby nematodes, and one nematodes-infected insect can produce hundreds of thousands of the same.

Oh, and you've probably noticed there are still plenty of mosquitos.

> The fact is that up to now the free society has not been good for the intellectual. It has neither accorded him a superior status to sustain his confidence nor made it easy for him to acquire an unquestioned sense of social usefulness.
> - Eric Hoffer -

Neglected science and neglected scientists

"Is there not room enough on the mountains . . ., that you should covet that which is but the drop and neglect the more weighty matter."
- Doctrine and Covenants 117:8 -

Humans usually confuse the "what-constitutes weighty matters" and what constitutes the "drops". We see all that goes on around us and think they are the weighty matters, often loosing sight of the fact that all the big obvious things rest upon vast numbers of a few little things that support the whole. Sky scrappers depend on bolts. Lions depend on good soil to produce the grass for grazing required by the zebras and antelopes which they consume.

Part of the reason for the confusion is that science seems to have done away with the idea of a creator, God, and so they don't see that the system is a designed system. Or if they do, they think it is an accidental design of infinite complexity, which of course is not a very scientific conclusion. The fact is the way the world works is infinitely complex. But it does work so whether you believe in God or not, they world is still there.

Humans often think that science is a weighty matter, not realizing it is basically the study of little drops. Each scientist specializes on a small field and advances knowledge and advances understanding in that area. His or her entire interest is in that tiny drop.

Occasionally discoveries within tiny drops turn out to be hugely significant and ends up turning the world upside down. The scientist, who is completely consumed by the tiny field and the small discovery, is usually surprised, and often dismayed, when the world is turned topsy turvy.

Most of the public thinks that science is the big weighty matters. They often do not understand that the large animals that are so impressive and magnificent depend entirely on a tiny invisible world which no one yet really understands.

Scientists with large grants tend to concentrate on "weighty matters". You know, things like disease, medicine, energy, technology, electronics. This is advantageous for independent scientists because it allows them to concentrate on the drops. You can be an independent scientist too.

Independent scientists

Yes, there are independent scientists. I am an independent scientist. I am not presently associated with any commercial or governmental laboratory. I choose my own research topics and conduct my own research. I am, therefore, independent of financial obligations and conduct my research using mostly my beer money.

This works for three reasons. First, I don't drink so that frees up a lot of cash that would otherwise go for consumption. But secondly, this allows me to do significant research in important but neglected areas of science. And third, while I am not independently wealthy, I have sufficient for my needs and find the activity of doing research pleasant and less stressful than golf.

Neglected Areas of Science

Yes, there are neglected areas of science. You might not think so based on the attention paid to science today. You read of massive research budgets, impressive numbers of publications, and the breathless predictions and claims of various scientists. "Tetra-di-hydro-benzylated-chicken wire MAY cure cancer. More research is needed."

Scientific policy is big news because it is supposedly policy based on science, which, of course, is never wrong. Except that science ideas keep changing based upon new things found out in science, so the policies are never any good. But it is good job security for the scientist.

Don't begrudge them that. Most people want job security. Look at teachers for example. They keep telling us that children need school with hardly any hard evidence that that is true. At least there is little evidence that children need the kind of schooling that schools provide. But it gives teachers job security.

But the truth is most scientists need to be paid so they can indulge their scientific interests. Scientists are people too, with families and mortgages. This leads them to do research that is valuable enough that someone will pay them to do it. It's like golf for businessmen. They must work at a business so they can afford to play golf. The dark side is that some scientists have to kill things, like mosquitoes or other humans, to make a living.

However, new discoveries with the possibility of further advances lead to scientists rushing on in new directions and leaving many gaps in our knowledge of routine matters of how the world works. For example, when the electron microscope was invented scientists rushed to use this powerful new tool to further our understanding of microscopic structure and activities. However, most of the world has never been adequately studied with the old, reliable, light microscope. In fact, most of the natural world hasn't been studied at all by any real scientific means.

In addition, science can sometimes be a set of blinders to understanding. The nature of science is to concentrate on narrow topics, questions, and areas of research. This enables scientists to find answers to complex albeit inconsequential questions. However, it also often blinds scientists to how the details interact in a wider view, and possible avenues of research that could be exploited.

For example, I spent many years studying nematodes from a variety of points of view. Nematodes are unsegmented round worms. Many of them are animal parasites while others are plant parasites. Many more are free-living, and we don't know much about how they benefit or harm the environment. I will not bore you with how I came into this study.

Then, one day, I attended a lecture by a person who was doing research on nematodes. It was a fascinating presentation and the discoveries in his lab were impressive. However, in having a discussion later, over light refreshments, he admitted he really knew nothing about nematodes, except the one experimental animal he was working on. He was a geneticist with little to no understanding of the role of nematodes in my garden and pasture. He had no idea of the commercial applications of his research beyond genetics, and maybe medicine.

Of course, his research is important. But the field of nematology is a neglected field and there is much an independent scientist could learn that might benefit the world if he or she could just find someone to pay the mortgage while they discovered it.

Science is a big subject

It didn't used to be that way. Oh, I suppose the subject was always big, but the practice has historically been small. At least until money got involved. I think there is a pattern there if someone is looking for a scientific study to pursue.

A lot of the discussion in this book will relate to biology, but that is only because I am a biologist, and my interests are mostly in that area. That does not mean that there are not unexplored topics, in the other sciences.

For examples:
- in chemistry there is a phenomenon called self-generating membranes. While known about for the last one-hundred or so years, little is still understood about their formation or applications.
- The geology and mineralogy of anthills is intriguing. Just what minerals are mined, from where, and why?
- Magnetisms effects, especially on living things such as honeybees and pigeons are still poorly understood outside of scientific and industrial uses.

Purpose of the "Simply Science Institute (S^2I)"*

". . . the four beasts . . . are figurative expressions, used by the Revelator, John, in describing heaven, the paradise of God, the happiness of man, and of beasts, and of creeping things,. . ; that which is spiritual being in the likeness of that which is temporal; and that which is temporal in the likeness of that which is spiritual; the spirit of man in the likeness of his person, as also the spirit of the beast, and every other creature which God has created."
Doctrine and Covenants 77:2

I formed the "Simply Science Institute" (S^2I) in 2010, during the final years of my working career at a university, to promote STEM education, create an avenue for continued research after retirement, and make what I do sound important. Hey, two out of three ain't bad.

Anyway, it has been the vehicle for an award-winning newspaper column on science, the publication of several books on science topics for the lay person, several attempt at combining science and artistic endeavors, and has kept me out of my wife's hair. It has also been a more-or-less endless hole into which to pour extra money and time.

I have found science more fun than golf, and I can waste just as much money on equipment. Further, I am not restricted by limitation of time and place allowing me to escape from the reality of paying the mortgage to the reality of science whenever I am free from paying the mortgage.

The present publication will explore the contributions of the insect world to the welfare of the humanity and the environment. Especially, I will look at insects as a source of energy and nutrients for the rest of the world. But I hope it will not be lost in the details that this is a miraculous, fine-tuned world that could not have been created by chance.

It seems to have been forgotten that anyone can be a scientist. Of course, if you want wealth and fame, you should be a banker or a lawyer because it is the rare scientist who achieves fame or wealth. Those that do usually end up doing so in administration and government. But at some point, in one's life, we each must ask, "how much is enough?" And at that point they begin to play golf or join a bowling league. But a masochistic few can become independent scientists.

You may find that your studies enable you to provide useful and interesting knowledge concerning fundamental parts of our world to your friends, relatives, and community. And you will become taller, thinner, and better looking. You might even turn the world upside down.

And through your example you may make new friends and inspire others to become amateur and independent scientists.

* "Simply Science" is not to be confused with "simple science", which doesn't exist. Scientists make sure that science is never simple to protect their monopoly on the subject. Simply Science is my little business of explaining science so that even scientists can understand it.

All things work together

"There is full-time employment for all simply in exploring. the world without destroying it, and by the time we begin to understand something of its marvelous richness and complexity, we'll also begin to see that it does have uses that we never suspected and that its main value is what comes to us directly from mere coexistence with living things - the impact on our minds and bodies, subtle and powerful, that goes far beyond the advantages of converting all things to cash and calories."
Hugh Nibley

Understanding

I love the complexity and beauty of our world. I don't understand it, but I love it. And the closer I look, the more fascinating it seems. I enjoy the beauty of great vistas and giant trees. But I have always been even more fascinated by the little world, the tiny things, the nearly invisible life that supports all the earth. None of the vistas and fascinating macroscopic world can exist without the complex, finely tuned, and flexible foundation of the world of creeping things. This book is about some of those things.

> **Understanding dirt is probability beyond human capabilities.**

For the most part, humans seem to not understand that "all things work together for your good." They question why there are mosquitoes, why there are flies, why there are diseases, and on and on. I find it ironic that we call a beneficial insect one that attacks and kills an insect that we see as not beneficial, disregarding the fact that if there was not a harmful insect there would be no insect that kills it.

It doesn't end with insects either. Humans have no idea how to make a complex ecosystem of any kind. The soil itself is a complex of living and non-living components with little understanding of the literally millions of living things interact. Bacteria feed on decaying organic matter and we don't know what all of the bacteria are either. Nematodes feed on bacteria, protozoa, other nematodes, and even arthropods, while arthropods feed on all of these and other arthropods. Indeed, understanding dirt is probability beyond human capabilities.

The bold and the beautiful
Humans love the vistas and the magnificence of the world. Humans are fascinated by the grace and strength of wild animals. Humans often cherish the companionship of animal friends such as dogs, cats, birds and other creatures, greats and small.

But it all rests on the presence of literally millions of different little creatures and plants, mostly invisible, that create the foundation of this fantastic earth. When one of these is lost an entire ecosystem can unravel and humans don't understand why. Nature is a marvelous balancing act built on top of a pyramid of tiny creeping things. It is one thing to suppose that a complex organism was created accidentally from random events and evolved into complexity. It is unfathomable that the tiny world of creeping things, obscure plants, and single celled life is balanced and supports the world of everyday life by accident.

*". . . for the pillars of the earth are the Lord's,
and he hath set the world upon them."
1 Samuel 2:8*

Working with God
The potential for working with God's creation is enormous and mostly unexplored. For example, the beneficial use of insects and all the creeping things far exceeds common definitions and approaches. Eradicating the creeping things because we don't understand their role is foolish beyond belief. Using insects to control other pests is a limited process. This is simply because there can never be as many predatory

animals as there are prey. Even if insects became acceptable as a food source for humans, the insects themselves must feed on something and raising enough of their food to grow enough insects to feed humanity without competing with other food stuffs is impossible.

However, insects represent an enormous potential source of biomass, waste, and enriched biological materials. If managed properly insects and their various byproducts could become a source of fertilizer, medicine, food, and many other beneficial products to help increase production of human food and other materials. We have not adequately explored how we might use insects, or most of the microscopic world, and their by-products to enhance human well-being.

> **All things work together for your good.**

"Search diligently, pray always, and be believing, and all things shall work together for your good, . . ."
Doctrine and Covenants 90:24

PART 1
CREEPING THINGS OVERVIEW

CHAPTER 1 - INTRODUCTION

*"And God made the beast of the earth after his kind, and cattle after their kind, and <u>everything that creepeth upon the earth</u> after his kind: and God saw that it was **good**."*
Genesis 1:25

Yes, even the creeping things of the earth were considered good by God. Children generally love creeping things. Well, anyway, I did when I was a child and I'm normal. It's adults that seem to have declared the small things of the world bad, or at least unimportant.

Perhaps it's that children and bugs inhabit the same space, close to the earth. Many children like to catch bugs and examine them, even keeping them as pets. Well, I did and I'm normal.

One of our sons used to catch flies in the house and carefully carry them outdoors rather than kill them. The other son was fascinated by pulling the legs and wings off creeping things. Both eventually outgrew their early behavior. Those who don't outgrow these inclinations usually come to a bad end. I didn't and I did.

There are those adult humans who inhabit a "middle earth", so to speak, an alternate universe or existence, who find the creeping things of the earth fascinating and important. Of course, most people think such individuals are wrong, or weird.

But these weird, childlike, adult humans have not lost the childlike love of living things, nor the tendency to want to see what makes them tick. Somewhere along the way they failed to learn to fear and to detest that which they don't understand. Driven by unhealthy curiosity, they tend to march to their own cricket castanets.

It is for these lost souls that this book is written. Frustrated by the misunderstanding by the larger populace, deprived of the education that could have provided much satisfaction, haunted by unsatisfied curiosity, discouraged from lack of resource or encouragement, these poor souls hide their interest and passion for the creeping things of the earth. While they labor away at mundane tasks to make a living, they secretly want to be a biologist.

It is commonly believed that one needs higher education to be a biologist. I admit that I took this route, but I have discovered that it is only necessary if you want to be the kind of biologist that helps anyone. It's obvious that one needs to know what they're doing if they are going to tell other people what is wrong with them medically. However, you can tell people what is wrong with them morally, psychologically, personally, and politically with no qualifications whatsoever.

Certain other activities where one's actions might affect other people also require specific education. But if you don't intend to be useful and are simply curious, the education does more to prepare you to write research grants than to study animals. The same thing is true of plants I suppose. I've never really studied plants myself unless you count cactus. Personally, now that I am retired, I just use my golf and beer money to fund my research. Since I don't drink or play golf that has been enough.

One of the advantages of studying creeping things is that no one, except maybe your spouse, cares much if you want to study cockroaches, isopods, centipedes, snail, earthworms, nematodes, tardigrades, or paramecium. Perhaps you will be lucky, and your spouse won't care; but don't count on it. Mine cares! And that has required some ingenuity on my part. Overall, though, it just adds to the challenge.

If you are curious and find creeping things interesting, you may waste at least as much time, and money, on learning about them as you can playing golf. All you really need is a love of nature, curiosity, imagination, ingenuity, patience, and the ability to read and think and you can be a biologist. And there are many biologists who lack many of these credentials and do just fine.

Poikilothermic

This next portion of the book is brought to you today by the word
POIKILOTHERMIC.
(Hey, it worked for Sesame Street.)

If you were to spontaneously write a list of ten animals, the chances are that eight out of ten of them would be animals with back bones: vertebrates. If you want to make this a social affair you and a friend could take turns thinking of animals. What if you have more than one friend? Hmmm, I am not sure what to do about that. I've never had that many friends. But the results would be about the same.

Anyway, I'm not sure what the big fascination with vertebrates is all about for humans. Vertebrates only make up five percent of all the different kinds of animals out there, so if you think animals are fascinating creatures, you are probably missing out on ninety five percent of all the fascination: the invertebrates. I know most people think invertebrates are kind of creepy, I guess that's because a lot of them are, well, creepy. That is, they creep and crawl around for mobility.

Invertebrates are also poikilothermic and aren't around much in the winter. I kind of miss them. Poikilothermic animals vary in temperature according to the temperature of their surroundings so when the environment gets too cold, they die. They die because the water, in their cells, freezes. Water expands when frozen and that ruptures their cell membranes. (Without a boundary of some kind, like a cell membrane, an object ceases to be anything in particular and becomes nothing in particular. The same can be said for an organism, a home, or a country.)

Honestly, not all poikilothermic animals die in winter. Some of them survive by seeking sheltered areas where they don't freeze. Desert scorpions burrow deep or follow natural fissures into the soil to depths of several feet. Of course, at those depths the temperature stays constantly above freezing. Some species of mosquitoes load up on sugars which acts as an antifreeze in their blood. They then hide away in sequestered areas such as under tree bark or the dashboard of your car.

You probably call such creatures cold-blooded. This can be misleading because cold-blooded has multiple meanings. Sometimes the term "cold-blooded" is used to denote a lack of feeling or empathy, as in the phrase "cold-blooded killer". I am not sure if cold-blooded animals have feelings or not, but there are warm blooded animals who act in especially cold-blooded ways. Some Humans come to mind. So, the word poikilothermic is a useful distinction between those animals that can somewhat regulate their internal body temperature and those that cannot.

> **Vertebrates only make up five percent of all the different kinds of animals out there, so if you think animals are fascinating creatures, you are probably missing out on ninety five percent of all the fascination: the invertebrates.**

And while all invertebrates are poikilotherms, not all vertebrates are warm blooded. Only mammals and birds are warm-blooded. Warm-blooded creatures are called homeotherms. However, homeotherms' temperatures will also vary with their surroundings if they are left out in their surroundings for long periods of time. It's just that they have mechanisms to alter their body temperature temporarily so that it takes longer for them to freeze. But once water in their cells freezes, you won't see homeotherms around much either.

Humans are different. While we are technically homeotherms, we are mostly just "*environmentaltherms*". Don't bother to look that word up because I just made it up. It means we mostly just change our environment to keep it as close to seventy-two degrees Fahrenheit as possible. However, we can generally last longer than a poikilotherm when left out in extreme environments.

You're probably wondering why I am telling you about poikilotherms when there might not be any around right now. It seemed appropriate to say something about the cold in January and I can't think of much of anything about the cold that I find interesting except the absence of poikilotherms. Maybe I'm just feeling nostalgic remembering the last cricket to die.

That and the fact that I just added my fifty-fourth invertebrate poem to my collection of poetry and wise sayings concerning poikilotherms. I'm kind of a purist in that I don't include any of my own poems, just those written by true poets and universally recognized wise men. You know, people like Ogden Nash and Lyle Lovett.

It's genuinely interesting how many of our talented and wise men have devoted poetic attention to the lowly poikilotherms. I find that almost as fascinating as the poikilotherms themselves. Aristotle observed, "Earthworms are the intestines of the soil." Marcus Aurelius once said, "That which is not good for the beehive cannot be good for the bees." So, take note that poikilotherms were cool even before it was cool to be poikilothermic.

Objects of Study

In fact, the biggest problem you will have if you want to study creeping things is deciding on an experimental animal. There are literally millions of different kinds of animals, and only about five percent have backbones. All the rest are basically creeping things. There are over a million species of insects alone, not including related groups such as crustaceans and spiders. That is a lot of variety.

To take just insects for example, the number of insects in the world is impossible to estimate with complete accuracy because of their variability in time, place, season, and my limited mobility. However, complete accuracy has never been one of my major goals. Apparently, it isn't for a lot of entomologists either because numerous experts estimate there are 10,000,000,000,000,000,000, or ten quintillion, insects on the earth at any given moment. I suspect they rounded up because they were seeking funding for their research.

By the way, entomology is the scientific name for studying insects, and entomologists are those who admit to such a perversion. Entomology is a more scientific sounding activity than "creepologist". This latter term leaves the door open for unfortunate interpretations. So, if you want to become public in your study of creeping things you may want to use terms like entomologist, protozoologist, heminthologist, and the like. When people don't understand they tend to be afraid of you.

Anyway, ten quintillion would be 200 million insects for each human on the planet! Of course, insects are much smaller than humans, but some experts have estimated that would still be 300 pounds of insects for every pound of humans. See, when you stop worrying too much about accuracy and being of use to people you can make cool estimates like that.

This may seem staggering or even impossible, but consider a study done on North Carolina where soil samples to a depth of 5 inches yielded a calculation that there were approximately 124 million animals per acre, of which 90 million were mites, 28 million were springtails, and 4.5 million were other insects. A similar study in Pennsylvania yielded figures of 425 million animals per acre, with 209 million mites, 119 million springtails, and 11 million other arthropods. There seems to be no end of creepy things on the earth.

> Ten quintillion insects would be 200 million insects for each human on the planet! Of course, insects are much smaller than humans, but some experts have estimated that would still be 300 pounds of insects for every pound of humans.

Three fourths of all animals belong to the arthropods, animals with jointed legs. The word "arth" is Latin for joint, although in this context it refers only to "pods", which is Latin for foot. As far as I know, one cannot smoke an "arth" which distinguishes them from other types of joints. The Arthropods contain the insects, and as mentioned, there are more than a million insect species in the world.

Consider honeybees, wasps, mosquitoes, lady bugs, grasshoppers, houseflies, cockroaches, butterflies, moths, and beetles. If that list of common creepers isn't sufficient, consider that there are 120,000 kinds of flies alone, and similar numbers of most other common critters.

Of course, there are many more creeping things than insects and arthropods. Think about earthworms, pill bugs, slugs, and snails; all common, though not insects. Still, how would you describe them if not "creepy". What do you really know about any of them? Then there are the nematodes, billions, and billions right below your feet. Tardigrades are ubiquitous and indestructible.

It is common to say that the only sure things in life are death and taxes. However, the creeping things are always with us, and I am not talking about the IRS. No, the creeping things are with us in our daily lives, living in our soil, homes, on our skin, in our body's interior, and even continuing with us after we die.

It isn't hard to find them, although it sometimes takes ingenuity and determination to find enough of them to study. Then, of course, you must hide the fact that you are studying them. And dispose of evidence. But I am getting ahead of myself. These are skills for later chapters.

God created a fantastic world of marvels. The great majority of living things are tiny and invisible. But without this invisible and unappreciated base of the food chain, most other, more visible living things, would perish. Lions, tigers, and bears depend for their existence on ants, grasshoppers, and nematodes.

This invisible world is full of wonder, complex behavior, brilliant colors, fascinating lives and a surprising amount of sex and violence. The intricacies of this more-or-less invisible world of creeping things supports pollination, food supply, waste cleanup, and population control for all so-called higher forms of life.

CHAPTER 2 - WHAT REALLY BUGS ME

"Any foolish boy can stamp on a beetle, but all the professors in the world cannot make a beetle."
- Arthur Schopenhauer -

You know what really bugs me? How little respect humans have for bugs. I guess I shouldn't be surprised. Humans don't have much respect for humans either. But just the expression "that bugs me" implies a great disrespect for bugs. We never say, "Boy, that really humans me!" I wonder if bugs do.

Geologists call our present time the age of man because of the massive impact humans have had on the planet. However, a great deal of our impact has been made in combating bugs. Consider all the energy, blood, and riches that have been poured into the war on bugs, and then tell me who's having the impact here?

It's even worse than it seems. We have altered the face of the planet trying to fight off a handful of bugs that irritate us when there are thousands more that actually benefit us. When did you last hear anything about one of the beneficial kinds? Some bugs do unimportant things like maintain normal ecological cycles, decompose human waste and dead bodies, or pollinate our crops. These things are important. Why aren't we farming them?

What about *Aceria malherbae*? That's a mite that lives on bindweed. Yes, I'm talking about THAT bindweed that grows all over my garden. I could use more *Aceria*. Or *Tyta luctuosa*? You know, the noctuid moth that eats bindweed. Scientists have found that *Tyta* overwinters in the area I live in, just apparently not in my garden.

Aceria malherbae

Working secretly, in an underground bunker somewhere, a dedicated team of deranged individuals befriend insects that are friends to humans. My goodness, that's more than most humans do for other humans. Oh, all right, they aren't really secret or in an underground bunker. The rest of that description is spot on though. You could be one of those who believe in beneficial insects if you want.

What's incredible is that normal looking people are the folks that have given us *Diorhabda carinulata, Hylobius transversovittatus* and *Jaapiella ivannikovi.* All of these bugs work tirelessly in our behalf killing off Tamarisk, Purple Loosestrife, and Russian Knapweed. Just because the average person doesn't know what any of those weeds are either, doesn't mean that there aren't bug people somewhere doing us a big favor.

Here's the big question though. If one of these *Diorhabda*'s gets into your house, would you say, "thank you" and place it gently back outside? No, most humans would call the local pest control company. They'd rather have an insecticide in their home, than insects. That's what really humans me!

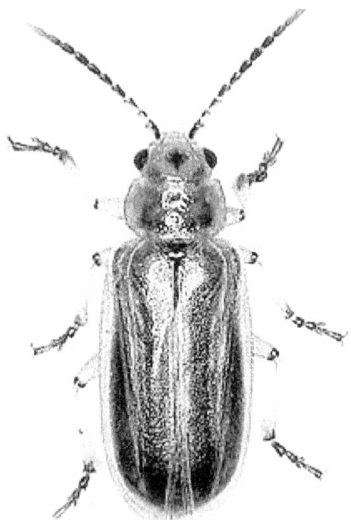

Diorhabda carinulata

I have buried the serious part of this column as deeply as possible into the narrative so as to not disturb the casual reader. However, this is where I must be totally honest and serious. Resistance to insecticides, herbicides, and even antibiotics in medicine may be one of the biggest problems facing mankind, assuming climate change doesn't get us first. The logical solution to a problem like Tamarisk is to use natural ecological pathways and methods, such as bio control, to live in balance with the things that really bug us.

At least let's have a little respect for *Tyta luctuosa*. Bindweed is killing my garden. Then the next time you see defoliated Tamarisk, have a moment of silence for *Diorhabda*. Show a little respect for the good bugs by learning more about your favorite creeping thing. That will accomplish greater good than most politicians.

> **Resistance to insecticides, herbicides, and even antibiotics in medicine may be one of the biggest.**

What's in it for you?
- Are you tired of fantasy sports leagues?
- Does golf give you more anxiety?
- Do you find most of television, and entertainment in general, crude, and distasteful?
- Are you looking for a meaningful activity that you can engage in that requires brains and dedication, as well as being crude and distasteful?

If so, you should probably consult the personal improvement section of the local library or bookstore. But if you are less fastidious about how you spend your time, and if you are fascinated with living things, especially those that creep and crawl upon the earth, then this book might be just the thing you need. In the following chapters I will endeavor to tell you a little about a few common creeping things, but in an uncommon way.

- Do you know where the Lady Beetle gets its name?
- Have you ever seen a pale-blue pill bug?
- Do you know what grasshopper tobacco juice is?
- Can cockroaches be drowned in the toilet?

In the following chapters I will tell you some of the odd things about various creepers, but more importantly I will point out the things we don't know so that you can go discover them yourself. I may make a few suggestions about how, and experiments that could be done, but the point is for you to figure that out for yourself. It is a lot easier for me to ask questions than go find out the answers myself. It is even easier to try and convince you to solve the problems than to solve them myself. You'll see.

You will be surprised how many approaches there are to solving questions about living things and how much we think we know, but don't. You may find you will need to consult others for help. Sometimes a fresh new perspective, and friend, can help. Be discriminating, however. Not everyone will view your interest favorably.

Sometimes specialized equipment may be needed that you can borrow or beg. Don't steal! That would be against the high ethics of invertebrate biologists. However, I am not sure how much one should depend on the ethics of biologists who study creatures that don't have a backbone.

Sharing

One problem you will want to plan for in advance. Once you know something other people don't know, there is a tremendous desire to want to share the information. It may be necessary for you to write papers or volunteer at local clubs to give talks. It's doubtful though because you will likely discover that your spouse and current friends aren't all that interested in how much detritus an earthworm consumes in a day.

New experiences await! Fame and fortune are just around the corner! Well, okay, I lie about the fortune, and the fame may be somewhat local . . . if that. But you will discover fascinating things about the world in which you live that will color everything you say and do. You spouse and friends will say that is exactly the problem.

You may discover that "creepologists" with actual college degrees are sometimes a little snobbish of amateurs. They have a lot invested in their own identity and will want to protect that. But perhaps just as many will be glad to listen, suggest and even aid and abet your activities. Those with research grants and obligations may not be interested. They simply don't have time for fun. Pity them.

> Once you know something other people don't know, there is a tremendous desire to want to share the information.

How to join the fun

In all honesty, if you are one of these people you probably don't need to read this book at all. Just grab a creeping thing and start making measurements and observations. But if you are paralyzed by the possible number of subjects and approaches, this book might help a little to narrow the field - both as to subject matter and method. If you are wise, it will convince you to abandon the effort entirely and join a group of Entomologists Anonymous (EA). I am told they have a wonderful twelve step program that I have never completed. Nice people though.

Bibliographies: To clarify, there are many ways of studying any subject. For example, one could begin simply by trying to collect a complete bibliography on what is known about an animal. This is no small task and can be an important contribution to your personal knowledge and to the world to have such collections developed. However, knowing everything already known is not necessary for asking your own questions or performing your own experiments. Knowing what is known often gets in the way of knowing new things. If you take this approach, try to find something that little is known about., It makes the task easier.

Collections: Collecting specimens from a region and/or of one species can be valuable to show variation in species and aid in understanding biogeography and evolution. Collections themselves can be artistic expressions and are highly informative to others who are interested. They are also useful to local agriculturists, garden centers, nature reserves, and pest control organizations to identify possible organisms in their region.

Illustrating: If you have an artistic bent, scientific illustration, drawing, painting, photography, and even sculpture of invertebrates is a neglected field. However, accurate illustrations and models are extremely valuable for several reasons. For example, they promote awareness and understanding, are educational tools, may illustrate differences and unusual features, and be a unique aesthetic contribution. Much of the early work by naturalists were the creation of detailed drawings.

As an extreme example, I once created fourteen orchestrations devoted to trying to depict musically several invertebrate organisms. Each musical piece is accompanied by a brief description and poem. This companion CD can be found either accompanying this book or separately at many music-download sites on the internet such as i-Tunes, Amazon, CD Baby, and others. It is entitled "Everything That Creepeth Upon the Earth" by Gary McCallister and his Flaming Moth Orchestra.

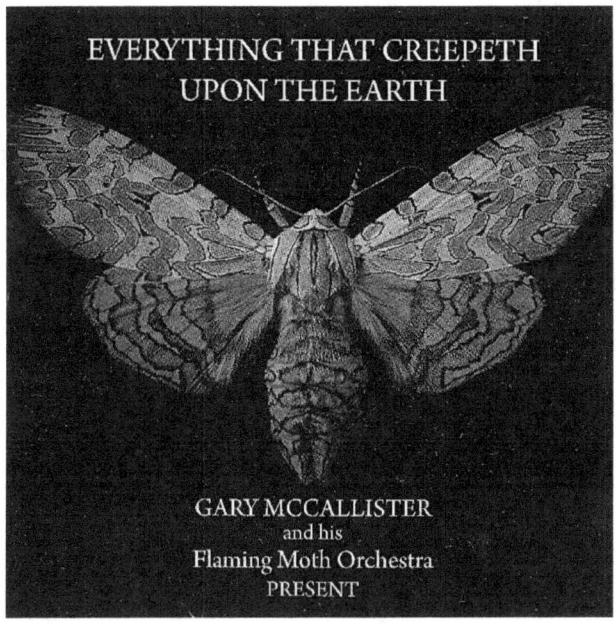

I introduction
two early morning beehive
333 when beauty is the beast
4our grasshopper on grass
v monarchs and milkweeds
six lady bug
7 mantis religiosa
8ight late night pill bugs
IX crickets and the quarter moon
10 Skeeter mountain
11ven the scarab of raw
xi luna
13 spider in firelight
4teen flaming moth
15 firefly

And, of course, there are numerous scientific approaches to studying any given organism.
- Biogeography and ecology are interesting ones for natural occurring organisms. Where do the organisms lie and what is their distribution over space and time? Are there different species in different locations? Why?
- How does a species affect the environment in space, time, and energy demands? How much food do they consume? What eats them? How much energy is in one individual? How many individuals are there in a given space?
- What do they eat? How much? How do they locate food? How do they eat? Can they see light? Do they respond to color? Can they hear? Do they sense vibrations? Are they directional? Can they sense temperature or moisture? How? Are they nocturnal?
- Do they get sick? Why or why not? How long does an individual live? How quickly do they decompose? Do they have other organisms living with, on, or in them?

- How many offspring do they produce? When? How often? How?
- What are they composed of? Can it be eaten? Nutritional value? How are the unique from other organisms in this way? And they be farmed for fish food or other resources such as chitin or calcium.
- What kind of heart, lungs, digestive, or nervous system do they have? How do they work? How are they different from other animals in this way?
- What are their requirements for breeding places, nesting sites, shelter, range, habitat, etc.? How do they respond to extremes and change?

> "If learning is not made public, it is a waste."
> Chaim Potok (The Chosen)

Obviously, there are way too many questions about way too many organisms. What I have found is that what is reported as the facts about a group of organisms is seldom true in detail about all similar organisms, and often the differences are simply not known.

For example, to read the literature you would think there is nothing new to learn about mosquitoes. However, there are approximately, as of this date, about 3,500 species of mosquitoes. One hundred seventy of those are found in the United States alone. However, no one really understands what the primary source of nectar is for male mosquitoes who do not take blood meals. In addition, there is growing evidence that each different species has uniquely different behaviors, such as the height at which they fly looking for mates and food, speed at which they fly, colors they may be attracted to, nature of saliva, which is the sources of the itching reactions, resting areas, preferable food sources, and so on ad infinitum.

In the following pages I will lay out for you some suggestions of easily obtainable and interesting invertebrates that are incompletely known about. I will introduce you to them generally and explain some of the reasons they may be important or interesting. At times I may make suggestions about where to find more information, or what kind of equipment might be useful. However, the point of doing science is not to "know", but to "do". Happy research!

CHAPTER 3 - A ROSE BY ANY OTHER NAME

"It ain't what they call you, it's what you answer to."
- W.C. Fields -

It occurs to me that I probably should say a word about the names of creeping things. That is a hurdle you might run into. Scientists do some funny things with names.

Without a classical education, there is something missing. It used to be that scientists came up with romantic and meaningful names for things scientific. Of course, that was back when everyone spoke Latin. So many organisms have Latin names with historical meanings. We still use Latin today to name living things, but the reasons are completely different.

In the old days, everyone spoke Latin, so it was the language everyone agreed upon. Now, no one speaks Latin, so it is the only language everyone can agree on. Since Latin isn't politically correct anywhere, it has been decided that Latin is politically correct everywhere. Very modern, don't you think?

Of course, today's scientists have probably never taken Latin. Does math or a year of high school Spanish count? Consequently, in a burst of creativity seldom seen, they have devised new pragmatic ways of naming molecules, enzymes, cells, and just about anything else you can think of.

The first fallback position after abandoning Latin was to use letters of the alphabet. So, we named proteins in the blood, according to the sequence in which they were discovered, labeled them a, b, c, d, etc. Eventually a daring young scientist broke ranks with this system also and began calling various items by number. While efficient, both methods lack emotion and whimsy.

I had always wanted to take the next step by discovering something new and giving it a unique name like Hazel, or Frank. Unfortunately, meteorologists beat me to that by naming tropical storms and hurricanes with such names. Anyway, I think there is something missing when we don't make up poetic names anymore.

<u>Biblical precedence</u>
Of course, many scientists do not take the story of the biblical creation seriously. However accurate it may be there are a couple of important points. One, of course, is where does spoken language comes from? No other animal has that gift to our degree. It comes from something unique.

The second thing is why do men need language and what is it used for? Apparently the first use of language that God gave to Adam, and no other animal, was to name the animals. And we are still at it. And each name creates a category of anything we may think that is like the thing named. The whole concept of classification began with language.

> *"And out of the ground the Lord formed every beast of the field, and every fowl of the air; and brought them unto Adam to see what he would call them: and whatsoever Adam called every living creature , that was the name thereof.*
> *Genesis 2:19*

But the job wasn't over because Adam was looking for a helpmate. So, God made Adam a mirror image and presented it to him. Of course, man, in his typical way, named her as if she were another animal and chose a name based upon categories, he had already created in naming the other animals. Adam called her 'woman" because she was like the category man. It was a biological name. She is not a person but a type. Notice that Adam did not speak to her. He spoke about her.

However, there is time and multiple events that take place before Adam gives his wife a name, Eve. This is not a classification or a category. This is the name of a person with individual will and attributes, unique from all others. Recognizing Eve as an individual human only occurs after mistakes have been made and God has instructed them both as to their roles in mortality and the significance of their relationship.

I mention all this to draw attention to the fact that when we name a scientific type, we are not naming an individual. Scientific types are not free to choose their fate, they are unaware of their mortality and time restrictions, and they are governed by instincts and built in behaviors.

Scientific names

My favorite scientific name is *Macrocanthoryhnchus hirudinaceous* because it is so much fun to say at parties. Well, I assume it would be fun to say at parties. I've never really had a chance to do it because I'm never invited to parties. If you are going to use *Macrocanthoryhnchus hirudinaceous* at your next party, practice first. Otherwise, you might spray on someone. Oh, and use it early in the evening. Things could go badly if you waited until after refreshments.

Macrocanthoryhnchus hirudinaceous

The name *Macrocanthoryhnchus hirudinaceous* was chosen by a guy named Pallus back in 1791. Biologists keep track of those things to avoid arguments later. Like when I wanted to name my son Chance. My wife said that if I did, she would name our daughter Community Chest. So rather than do either, we named him something completely different. (Sorry Zane.)

> I think there is something missing when we don't make up poetic names anymore.

You might be surprised to know that there are rules about these things. I'm not talking about naming human babies. You can name them just about anything you want: a, b, c, 1, 2, 3. No, naming plants and animals is far more important. If you mix up a couple of kids, it's no big deal. You can always make more. But if an animal were to become extinct, you would never know what to call the animal that wasn't there anymore. So, rules are important.

All animals and plants have two names because Linnaeus said they must. He's the guy who sort of invented the binomial system of nomenclature which just means that all plants and animals have two names. Hmmm, I seem to be caught in some kind of endless loop here. . . . My major professor once discovered a new species of tapeworm in a shark when he caught a shark by accident. He named it *Serendipitous serendip*.

Anyway, *Macrocanthorhynchus* literally means the worm with very large hooks on its big nose. I have seen some humans that have rings in their noses. If they have very large rings, they could be called *Macroannulorhynchus*. If they are only small rings, I suppose they could be called *Microannulorhynchus*. Isn't this fun? Of course, we know the macro and micro parts refer to the rings because they precede the ring instead of the rhynchus. A ring in a large nose would be a *Macrorhynchusannulus*. See what a classical education can do for one?

CHAPTER 4 - SIGNIFICANCE OF CREEPING THINGS

"Even these of them ye may eat; the locust after his kind, and the bald locust after his kind, and the beetle after his kind, and the grasshopper after his kind."
Leviticus 11:22

    The millions of insects, not including related groups such as crustaceans and spiders, consume food for energy, produce waste in some form, die and decay completing the carbon cycle, and enrich the earth. Many insects are herbivores and attack our crops, many more are herbivores that for various reasons are not economic competitors with humans. Large numbers of insects are carnivores, mostly preying on other insects, sometimes to our benefit. And perhaps the greatest number of insects, although of the mostly smaller varieties, are decomposers helping reduce waste back to materials beneficial to the environment and humans.
    Most research in entomology is directed at those insect species that are either harmful, or those that are beneficial, to humans in some way. Whether something is harmful or not depends on one's definition of harm. Entomophobia, the fear of insects, causes car accidents, but that does not mean that a moth in the car is a harmful insect. It means humans in general have completely lost perspective and are not in touch with the earth.
    Most sources state that there are approximately thirty-eight (38) harmful insects. Of course, such a list might include aphids and there are several hundred species of aphids. Perhaps more. I suspect most sources just count all the aphids as one harmful insect. Still, the number of harmful insects is surely not an impressively long list.
    The list of "beneficial insects is equally as brief and just as obscure for the same reasons. Lady bugs do eat aphids, and there are many kinds of lady bugs. But not all are equally beneficial at eating all kinds of aphids and some eat more than others.
    More research is desperately needed on how to encourage and promote the growth and success of the various

pollinators. However, very few insects have been successfully raised in captivity, and pollinators are especially difficult because of their dependence on other living things to survive. One cannot successfully raise pollinators without their food source which is in some way tied to the success of some plant. The plants success, in turn, is tied to the success of the pollinator. Such examples of "all things working for our good" seem to be beyond human understanding.

> **The millions of insects, not including related groups such as crustaceans and spiders, consume food for energy, produce waste in some form, die and decay completing the carbon cycle and enriching the earth. Does anyone have any real idea how much this benefit humanity?**

Human food

I find it interesting that some people don't want insects in their houses or gardens, but don't mind eating them. In recent years there has been added attention to using insects as animal food. Insects reproduce rapidly and many can live off material that is not for human consumption. Black Soldier Flies can live off almost any decomposing material, even feces, and then be used as a source of protein for other animals. This would allow humans to raise a protein source to feed to our commercial animals without competing with the human food supply.

This already occurs on small scales by raising insects to feed to pets such as reptiles, amphibians, and fish. There has been increasing interest in raising insects on a larger scale for the use of feed stock for domestic livestock such as cattle, sheep, pigs, fish, and poultry. Some large industrial scale projects are already underway in Europe.

There has been increasing interest in using insects for human consumption. Humans have always eaten insects for survival and in some cultures, it is a common occurrence. It is not worldwide however, and the resistance to eating insects is difficult to overcome. However, in recent years there have been commercial operations to create cricket flour and other such products to add to normal food stuffs.

All things work together - Reprise

"For the most part, humans seem to not understand that "all things work together for your good." They question why there are mosquitoes, why there are flies, why there are diseases, and on and on. I find it ironic that we call a beneficial insect one that attacks and kills an insect that we see as not beneficial, disregarding the fact that if there was not a harmful insect there would be no insect that kills it." (First paragraph of this publication)

The idea that all things work together for our own good is so true it is almost as if it were part of some grand design. Some believe that because they can invent a story about how it all happens accidentally that their story proves that the design idea is false. But the magnificence and complexity of the little world sort of makes it difficult to believe in the theory of accidental creation.

The potential for the beneficial use of insects far exceeds these common definitions and approaches. Using insects to control other pests is a limited process. This is simply because there can never be as many predator animals as there are prey. Even if insects became acceptable as a food source for humans, the insects themselves must feed on something and raising enough of their food to grow enough insects to feed humanity without competing with other food stuffs is probably impossible.

However, insects represent an enormous potential source of biomass, waste, and enriched biological materials. If managed properly insects and their various byproducts could become a source of fertilizer and by products to help increase production of human food and other materials. We have not adequately explored how we might use insects and their by-products to enhance human's well-being.

> **Insects represent an enormous potential source of biomass, waste, and enriched biological materials. If managed properly insects and their various byproducts could become a source of fertilizer and by products to help increase production of human food and other materials.**

Life cycles

With over a million species the life cycle for most is still not known in any detail. The bionomics of their entire life cycle, possible food sources, temperature and humidity requirements and time required for their life cycle are incompletely known for all except the few that are parasites or important domestic or economic pests. Even for those important to man we often do not have a full understanding of how many generations a year might be possible or even how many realistically exist.

What is known generally is that insects reproduce much more rapidly and have shorter life spans than vertebrates. This means that at any given moment there are not only 300 pounds of insects per pound of human, but that shortly there will be 300 more pounds of a new generation of insects. This is not a weight; it is an avalanche.

If we assume four generations per year, then there are, annually, 1200 pounds of insects per pound of human. Whatever happens to all those dead insects? Well, of course they decompose and are recycled in the environment. Twelve hundred pounds of insect per year times ten quintillion is an exceptionally large number indeed. That is the approximate amount of dead and decaying insects that are each year decomposing and becoming additional compost for the earth.

Do we thank them?

> **Twelve hundred pounds of insect per year times ten quintillion is an exceptionally large number. That is the approximate amount of dead and decaying insects that are each year decomposing and becoming additional compost for the earth.**

Food and Energy Sources of Insects

The food consumed by insects is also imprecisely known, by source or quantity. We do understand that they are not plants and are therefore secondary consumers, herbivores, or carnivores. There are both, but they all derive their energy from other biological material. While some of those food sources may compete with humans, many others derive their energy from waste material as decomposers and scavengers. These activities are a great service to mankind. For example, at 1200 pounds of dead insects per pound of human annually, it would not be long before humans were buried in dead insects.

While there are numerous insects that compete with humans for our food supply, there are many more that feed on decomposing plants, animal remains, and waste products. As this happens in a distributive manner and at an almost invisible scale in unseen places, it is not generally appreciated. The following are areas where insects especially, are beneficial to humanity in ways that are seldom considered.

Waste products

Like all secondary consumers, insects produce biological waste material. Insect waste is often referred to as "frass" or sometimes as "castings". This waste contains considerable unused energy, as it does in the manure of larger animals such as vertebrates.

Animal waste had been used since the beginning of agriculture to enhance plant development. Only in the past fifty-or-so years has mined sources of nitrogen, potassium, and phosphate been substituted for animal waste as plant fertilizer and soil enhancers.

Because insects are small and often unrecognized in the environment it has not been well established what effect their waste material might be contributing to the natural ecology, or the potential they may hold for crop enhancement by using them as bio-fertilizers. Considering the massive numbers and biomass that insects represent, the amount of animal waste product must be impressive. Once again it probably represents a huge amount of animal waste available for replenishing the soil.

Chitin

Arthropods in general, and insects in particular, have an additional unique feature: an exoskeleton. The exoskeleton of arthropods varies in detail from species to species, but they all contain some quantity and form of a material called "chitin".

Chitin ($C_8H_{13}O_5N$) is a long-chain polymer of N-acetylglucosamine and is a derivative of glucose. It is a primary component of cell walls in fungi, the exoskeletons of arthropods, such as crustaceans and insects, the radulae of mollusks, cephalopod beaks, and the scales of fish. The structure of chitin is comparable to another polysaccharide – cellulose as found in plant cell walls. It can also be compared to the protein keratin.

Chitin has proved useful for several medicinal, industrial, and biotechnological purposes. Plants have receptors that can cause a response to chitin. When the plants chitin receptors are activated by chitin, genes related to plant defense are activated. In turn, chitin has also been found to be an insect disruptor.

Chitin also has some promise in medical applications and as a possible biodegradable plastic. For uses in these areas the chitin is usually broken down into a product called chitosan. The most economical sources for chitin for chitosan conversion has been using the waste products of the shellfish industry such as shrimp, lobster, crabs, and such. The use of raw insect chitin has not been adequately investigated.

The following table shows the amount of various plant growth compounds found in marine chitin from food processing plants, mostly from shrimp.

Waste/parameter	pH	C(%)	N(%)	P(%)	K(%)	C/N
	8.55	28	4.98	1.42	0.05	5.62

The figure below shows some of the ways in which chitin has been shown to be useful in many areas of science but especially in food production. There has been increasing interest in using insects as human food. Perhaps more realistic is the use of insects as animal feed. However, one area that is somewhat overlooked is that of using insects as a fertilizer.

Besides the nutrient factors that are obvious, chitin has shown to have positive effects on the soil and on various crops, even to stimulating resistance against insect pests. It's value as a plant growth additive/stimulator is currently being researched in numerous parts of the world. Want to help?

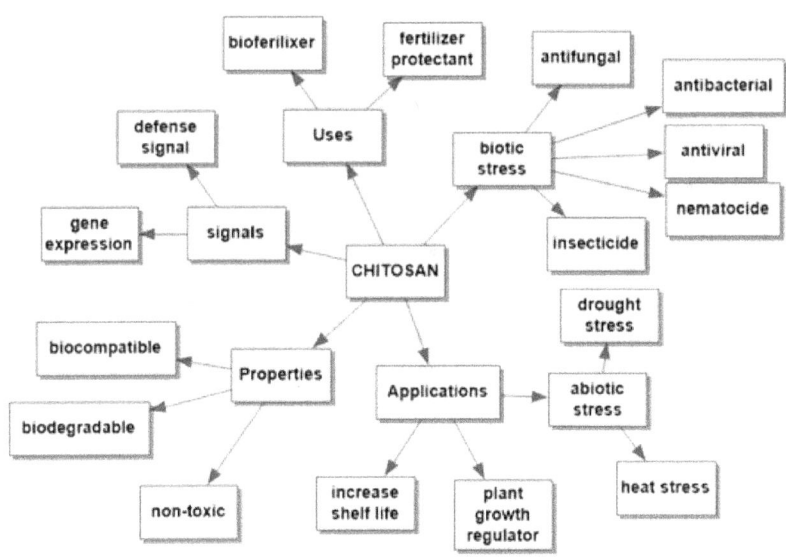

Arthropds as fertilizer

Because of the unique properties of insects as discussed above, it seems like they must contribute a significant, if dispersed, benefit to the natural ecosystem. The combined contribution of the insect, not to mention other arthropods, housecleaning, biomass, frass, rapid reproduction, and total enrichment of the environment and soil for plant growth must be significant, if unexplored.

Summary

The final illustration below attempts to bring together in one place some of the contributions that creeping things make to the benefit of mankind. Most people may be familiar with the some of the agaricultural contriobutions, but may be surprised at the medical applications. This may help guide your own choices in research areas.

> The combined contribution of insects, not to mention other arthropods, in housecleaning, biomass, frass, rapid reproduction, and total enrichment of the environment and soil for plant growth is significant and unexplored.

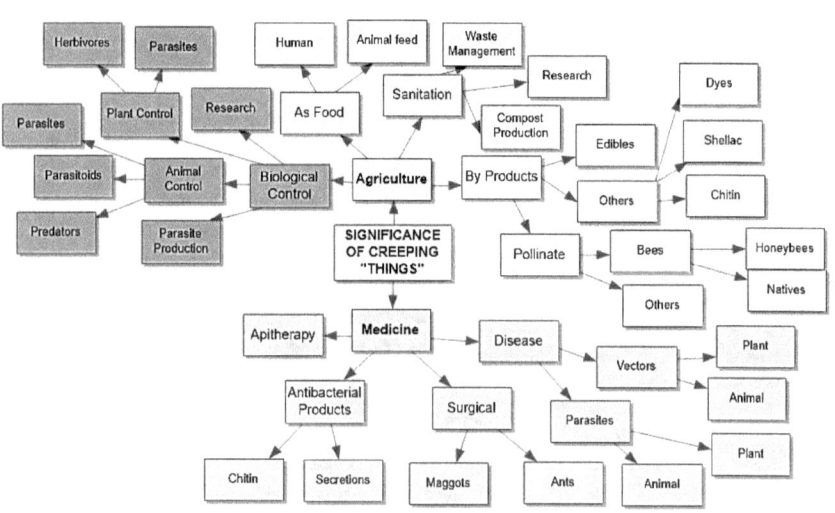

"And every creeping thing that flieth is unclean unto you: they shall not be eaten.
Deuteronomy 14:19

CHAPTER 5 - NEGLECTED SCIENTISTS

*There is no appeal from the law of nature.
It was made for beast and bird and creeping things.
Will the human never learn that in the eye of the law
he is not different from the things that creep?
John Charles Van Dyke*

John Ray may be one of the most important biologists you have never heard of. I guess that may not be very significant since most people couldn't name a single biologist if they tried. Heck, a lot of people know a biologist and don't even realize it. They probably just think that person is kind of weird or something. It's not like we are famous athletes or something.

Anyway, John Ray was an English naturalist who wrote about Botany, Zoology and Natural Theology. Natural theology is the study of science to prove the existence of God. Don't be too shocked, most early scientists did science for this very reason. In fact, most of the fundamental discoveries of modern science were made by believing Christians. And he was an "early scientist" living in the late 1600's.

He was a professor at Cambridge University but lost his position because he would not swear an oath of allegiance to Charles II. This was due to his puritan religion. After this he started on a journey, first through England and Wales, then later on to Europe in an attempt to describe and name all the living things, plant, and animal.

This effort predated Carlos Linnaeus by fifty years and his work was heavily relied on by Linnaeus in writing his own *Systema Natura.*

Sure, everyone's heard of people like Linnaeus and Robert Hooke. Hooke was the irascible old scientist of the 17th century who discovered almost everything before anyone else could.

But have you ever heard of Olavi Sotavalta? Of course not. That's because he was a neglected scientist working in neglected scientific fields. But his contribution is still significant.

Olavi Sotavalta,
A Finnish scientist who studied insect sound

To be "socially useful" he became an entomologist. However, like many scientists he took refuge from the dry, unemotional practice of science in music. He is said to have had a beautiful singing voice, and, more importantly, he seemed to have perfect pitch.

Okay, I admit Robert Hooke first reported that he could identify ". . . how many strokes a fly makes with her wings (those flies that hum in their flying) by the note that it answers to in musique during their flying." This was reported by Samuel Pepys, a British civil servant who was a friend of Hooke's.

The detailed private diary that Pepys kept from 1660 until 1669 was first published in the 19th century and is one of the most important primary sources for the English Restoration period. It provides a combination of personal revelation and eyewitness accounts of great events, such as the Great Plague of London, the Second Dutch War, and the Great Fire of London. As such, it is perhaps a classic example of an independent scientist.

Anyway, many years later, in 1952, Olavi used his perfect pitch to identify wingbeat observations on flying insects in free flight. His peculiar approach shows what can happen when scientific insights emerge when separate disciplines collide. That's most likely to happen with independent, amateur scientists.

Sotavalta's papers can be found in the National Library of Finland. They are curious mixture of letters, monographs, and stacks of sheet music. Many of his compositions are named for birds or insects.

Perhaps the strangest of Sotavalta's papers documents the songs of two nightingales. Sotavalta heard them during successive summers while staying at his summer house in Lempäälä. In the paper it eventually become clear that he is trying to apply music theory to birdsong. "The song of the two Sprosser nightingales (*Luscinia luscinia* L.) occurring in two successive years was recorded acoustically and presented with conventional stave notation," he wrote.

But the relevant paper to this tale is his paper in Nature where he describes the uses of his "acoustic method" of identifying insects using his absolute pitch. In it he theorizes about insect wingbeat: how much energy it consumes, and how it varies according to air pressure and body size. Of course, scientists, as usual, have made it far more complicated by applying laser beams to wing beats to do research on malaria carrying mosquitoes and crop pests.

I share this story to introduce what is to come; a series of papers about insects to suggest and encourage independent scientists to pursue knowledge for the fun of it. Humans remain deeply dependent on pollinators like bees, moths, and butterflies. However, a 2016 report showed that 40 per cent of invertebrate pollinator species are under threat of extinction. It's because of this love-hate relationship with insects that we urgently need better ways of tracking different species - better ways to differentiate between the bugs that help us and the bugs that hurt us. Perhaps even ways to discover that all things work together for our own good.

There are other creeping things that beg for understanding, but they will be covered in other Simply Science contributions. For now, insects lend themselves to the independent, amateur scientists.

And we know that all things work together for good to them that love God, to them who are the called according to his purpose.
Romans 8:28

Search diligently, pray always, and be believing, and all things shall work together for your good...
Doctrine & Covenants 90:24

> **Have you ever heard of Olavi Sotavalta? Of course not! That's because he was a neglected scientist working in neglected scientific fields. But his contribution is still significant.**

PART 2 - THE CREEPERS

Insects and flowers

The insects are dying. Its fall and it's turning cold. The Honeybees in my hive are balled up at night now, and there isn't much foraging, even on the sunny days. I see spiders laying their eggs on the side of the house. A Daddy Long Legs, technically not a true spider, hangs lethargically by the front door. The Praying Mantises are big, and fat, and slow. The word Mantis means prophet and so I assume this foretells the coming of winter.

The garden is dead. Only wilted and discolored flowers remain in most places. Fruit has been set, seeds have been shed, and nuts are in the shell. I still need to clean out the old growth in the garden, but I don't feel any hurry. Fall is for slowing a little, taking one's time, and feeling a little glad that the work is over, but feeling a little sad that the growing is over.

Sometimes the most important truth can be hidden in plain sight. There are over 250,000 flowering plants that have been described. That is probably a modest estimate, but I am not a Botanist and don't want to over-sell. There are over 750,000 insects described. That number is actually much bigger and is expected to go over a million.

Together this means that two thirds of all life forms are monopolized by these two groups. This is not an accident. These two groups of living things live together in an intimate way. Flowering plants could not exist without the service of insects to aid them in sexual reproduction, which we call pollination. And most insects could not exist without the shelter, surface, and food (nectar, pollen, and plant parts) provided by the plants. These two groups are completely symbiotic: dependent on living together.

This concept of living together is a delicate and changing arrangement. There are flowers like *Passiflora incarnata,* the Maypop, common in the southern United States in areas like Tennessee, that are only pollinated by *Xylocopa virginica*, a carpenter bee. If the bee is lost, the flower will also become extinct. Or the "bearclaw poppy", *Arctomecon humilis*, which is only pollinated by a solitary bee, named *Perdita meconis*, unknown until just a few years ago. If the flower is lost the bee will go extinct. These last two live near the Virgin River in Southwest Utah, or Northwest Arizona, as you see it.

Sometimes this balance between organisms is upset and we call the result predation, or parasitism, or disease, or extinction, or pollution or some other term. The problem is that it is very difficult to know what will upset the balance between any two or three organisms. How do we know what to avoid, or how to avoid it? It is akin to a complex structure built out of toothpicks. It is hard to predict which toothpick can be removed and which cannot without causing the collapse of the whole system. Generally, humans don't have a clue what we are doing in this regard.

Mankind has put a lot of energy into killing insects. Many insects compete with us for our food. Some insects transmit diseases. But ironically, mankind relies heavily on the flowering plants for food and fiber. High mountain peaches, cherries, apples, pears, and apricots are just a few of the hundreds of plants we find desirable that rely on insects. So, if plants need insects, and insects need plants, and man needs plants, then doesn't man need insects?

The rest of this book discusses some little-known facts about a few common species of insects. I hope you enjoy.

> **Sometimes the most important truths can be hidden in plain sight.**

CHAPTER 6 - BLATTIDAE (COCKROACHES)

Phylum: Arthropoda
Class: Insecta
Order: Blattodea
Family: Blattidae (Latreille, 1810)

". . . a single cockroach will completely wreck the appeal of a bowl of cherries, but a cherry will do nothing at all for a bowl of cockroaches."
Paul Rozin

It's a little known, and even less appreciated, fact that I am probably one of the leading experts on cockroach biology in the western United States. I say this in all humility. I mean, could anyone ever say something like that in anything other than "all humility"? I suppose there could be one or two people in the state who are as knowledgeable about cockroaches as I am. They're probably mostly in the legislature.

I didn't set out to accomplish such an honor. I was going to become the first person to complete the life cycle of a parasitic nematode in vitro. A nematode is a worm and "in vitro" is a fancy way to say, "in artificial conditions". Parasitic nematodes can only be grown in a living host which makes it difficult to fully understand their complete biology.

Understanding their complete biology is necessary to identify the weaknesses in their life cycle that we could exploit to help control them. Some scientists like to talk about eradicating them, but that's just crazy talk. I'd be happy with prediction and control.

No one has completed the life cycle of a parasitic worm in a test tube to this day. This might not seem like such a high goal until you realize that over two billion people in the world suffer disease from parasitic worms. In fact, when surveys are done, even one third of the people in the United States test positive for intestinal worms. However, they are routinely misdiagnosed and unrecognized in medical practice.

It is even worse if you are a cockroach. Having examined several thousand cockroaches by autopsy I can tell you the cockroach doesn't make it to adulthood without becoming infected with either *Thelastoma buhoesi* or *Hammerschidtiell diesingi,* and often both. Take some satisfaction in that knowledge, although it doesn't really appear that the worms harm them all that much.

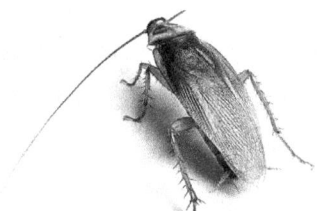

Periplaneta americana

But I digress. I am sure you are far more interested in learning how I came to my extensive knowledge concerning cockroaches. Well, I was supposed to complete a doctoral degree under Dr. Stanley Leland at Kansas State University on nematode parasites in cattle. Except he didn't get his grant that fall so I signed a contract to begin teaching at a community college.

Still, being interested in parasitic nematodes, I looked around for an inexpensive host animal. Did you know it costs a lot of money to buy a cow? I knew hamburger was expensive, but have you priced it on the hoof? I told my wife we could butcher one when I was through infecting it, but she wasn't very enthusiastic about that. In fact, she said, "Heck no!", or something like that.

Then, through a general biology lab, I discovered that cockroaches have nematode parasites, and cockroaches could literally live fine on the leftovers of my lunch. My wife said that was fine with her if I fixed my own lunch, never brought my research home, never spoke of my research in her presence, showered and changed clothes in the shed before coming in the house after work and one or two other minor stipulations. I was set.

There was a little problem at home when my wife learned I had graduated from a worm parasite of sheep to a worm parasite of cockroaches. She seemed to think I should be progressing in a different direction. And then there was the aspect of touching unclean things.

*"Wherefore come out from among them,
and be ye separate, saith the Lord,
and touch not the unclean thing;
and I will receive you, . . ."*
2 Corinthians 6:17

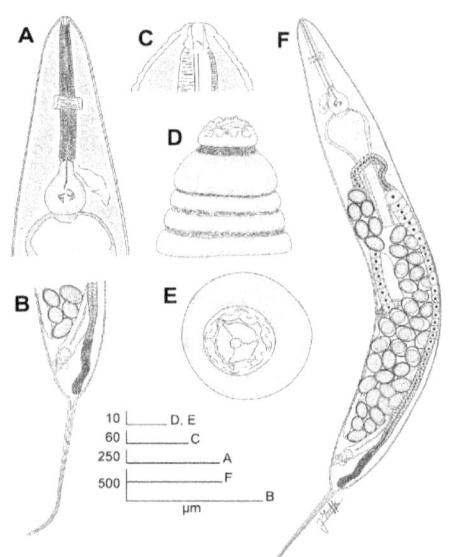

*Original illustration of Thelastoma bulhoesi
by de Magalhaes, 1900*

The surprising thing was that I needed to learn almost as much about the host animal as I had to learn about the parasite. And oh, the things I learned. For example, did you know that a cockroach can live up to a week with no head? Then it dies from dehydration, not starvation.

You can't really drown a cockroach. They can hold their breath for up to forty minutes. Hey, if there is a lot of interest, I could be persuaded to tell you about the experiments I did on the effect that swimming as exercise had on parasite infections in cockroaches!

I did discover that cockroaches can survive temperatures as low as 32°F for over thirty minutes, but their nematode parasites cannot. So, I could cure my roaches of their nematodes by chilling them in the fridge. This is extremely important to the cockroach ranching industry as it provides them with an efficient way of treating their herd.

Well, if there seems to be an unusual number of errors and typos in this chapter, I apologize. My wife refused to proof-read this chapter.

> **The surprising thing was that I needed to learn almost as much about the host animal as I had to learn about the parasite. It is almost always the case in science that one has to learn a lot about something other than what they are studying to study it.**

<u>The real one percent</u>

There are nearly 5000 species of cockroach identified on the earth so far. Only about thirty cockroach's species cause humans any grief at all, which is less than one percent of the known species. Maybe the others would cause us problems if they could, we just don't know. If we lived on Mount Everest, *Eupolyphaga everestiana,* a montane specialist that lives on Mount Everest at well over 16,400 ft (5,000 m) above sea level might bug us if we lived there also.

There are almost as many species of cockroaches as there are known species of mammals. Yet we don't designate all mammals as disgusting vermin because of a few mice and rats? Well, okay, perhaps that is a debatable point, but dismissing ninety-nine percent of all cockroaches on the basis of the few in your kitchen seems just as unfair as designating all humans bad because some of them are disgusting.

If you think you have a problem, consider leaf cutter ants of north America. Their nests are infested by the smallest cockroach on record. It's called an "ant cockroach". Well, duh. It is only a few millimeters long and is dwarfed by their hosts.

And you don't have to deal with *Megaloblatta blaberoidesboasts*, seriously, and has a wingspan of seven inches. These are only found in central and south America, thankfully you might add. Besides being one of the largest insects known they also can stridulate; that is make a sound by shaking their terminal abdominal segments. Instead of sounding like a cricket, however, they sound almost like a rattle snake rattle.

Most cockroaches have taken on ecological roles as decomposers. By feeding on decaying organic matter, they make nutrients available to other organisms. Consider your kitchen as just an environment and the cockroaches are enriching it for yet others of God's creatures.

Almost everyone has heard the catchy Spanish folk song, "La Cucaracha". But almost no one knows what it is about unless you speak Spanish. It is about a cockroach unable to walk because he has lost one of his six legs. Children seem to love this song, which may say something about children, or humanity. And almost everyone thinks it is a fun sing-along. But almost no one likes cockroaches.

"La Cucaracha"
The cockroach, the cockroach,
Can't walk anymore
Because it doesn't have,
Because it's missing
Two little back legs

People they say the cockroach
Is a very small animal
And when it gets into a house,
It'll soon be the Master of It all.

Of course, those cockroaches that do invade our homes are usual dwellers in unclean places and transfer potential pathogens into our lives. Perhaps more importantly, in many instances can cause serious allergies and asthma problems, especially in children.

What may not be as widely known is the fact that cockroaches are a very interesting and resilient pest that exhibits some very odd behavior and survival tactics. For example, cockroaches spend 75% of their time resting and can withstand temperatures as cold as 32° F. Here is a collection of other verified and fascinating facts about cockroaches.

- A cockroach can live for a week without its head.
- A cockroach can hold its breath for 40 minutes.
- Cockroaches can run up to three miles in an hour, which means they can spread germs and bacteria throughout a home very quickly.
- German cockroaches become adults in as little as thirty-six days.
- A one-day-old baby cockroach, which is about the size of a speck of dust and can run almost as fast as its parents.
- The American cockroach has shown a marked attraction to alcoholic beverages, especially beer.
- There are more than 5,000 species of cockroaches worldwide.
- Because they are cold-blooded insects, cockroaches can live without food for one month, but will only survive one week without water.

One final note about cockroaches that pertains to most, perhaps all, insects. Their nervous system is unlike ours in an incredibly significant way.

> **Old Testament**
> **Judges 13:4**
> Now therefore beware, I pray thee, and drink not wine nor strong drink, and eat not any unclean thing:

To illustrate, imagine you have built a small robot car that is powered by two electric motors, one to each rear wheel. If the right motor revolves more rapidly than the left motor, the car will veer to the left. If the left motor is faster than the right the car will turn right.

Imagine this car has two light sensors on the front of the car, set several inches apart. These light sensors are connected to the motors of the car and control the power to the electric motors such that the more light that hits the sensor the faster the motor turns. The right sensor is connected to the right motor, and the left sensor is connected to the left motor.

Imagine we have placed this car in a darkened gymnasium. It will not move because there is no light. But we have placed a remote-controlled light bulb in the center of the floor. When we turn the light on the car will begin to move. However, because of the distance between the two sensors, the amount of light striking the right sensor will be greater than the amount of light striking the left sensor. This will cause the right motor to revolve faster, and the car will veer away from the light until it is exactly facing away from the light so that the amount of light to each sensor is equal. It will also go as far away from the light as possible until the sensors are no longer stimulated.

Imagine you are observing this with a friend from the rafters of the gym. Your friend might say something like, "Wow, that thing really doesn't like the light. It runs and hides. How did you make it do that?" Of course, it doesn't "like" or "dislike" anything. It's a robot. It just appears to be a little like a cockroach.

Stay with me here. This is very applicable to you.

Imagine you make one small change in your robot; you connect the right sensor to the left motor and the left sensor to the right motor. Then you turn the light off, reposition your robot in the gym, and you resume your perch in the rafters.

When you turn on the light the robot moves, but this time it turns toward the light because the sensor on one side drives the motor on the opposite side. Your friend says, "Oh look, it likes the light and is moving towards it." But wait, something is drastically wrong. As the robot gets closer and closer to the light, each sensor gets more light, and this makes each motor go faster. The robot hurtles directly at the light with increasing speed. You friend screams, "Look out! It's attacking!" as the robot hurtles into the light demolishing light and robot in one grand violent act. "Wow!" Your friend observes after a stunned silence. "That robot really hates the light."

Imagine you painstakingly reassemble your robot. This time you add one more tiny change: a governor on the light sensors so that it increases speed until a certain light intensity is reached. Above that intensity the robot turns off the motor it is wired to.

Meanwhile, back in the gym, this time the robot turns towards the light and rushes towards it as before, but as it gets close it slows, and stops, and sits staring adoringly at the light bulb, never moving a motor. Your friend observes, "Oh look, it's in love with the light."

What has this got to do with you and me? Maybe nothing. But cockroaches, and most insects, have brains that connect to the same side of the body (muscles as motors) as their sensors, whether those are eyes or antennae. You and I have crossed nervous systems. The left brain controls the right side of the body and the other way around. Does that partly explain human aggression? And is the difference between love and violence a simple breakdown of the speed governor, the braking system? If you think these ideas are intriguing you would enjoy the book, "Vehicles: experiments in synthetic psychology" by Valentino Braitenberg, available from MIT Press and on Amazon. But please finish this one first.

THE DIFERENCE
There are wires from the eye
To the motor on the same side
And while the eyes see
It's the motor that guides.

If the eye sees on the right
The motor on the right drives
If the motor goes fast
The organism hides

If there are wires from the eye
To the motor on the other side
Then the eyes will see
But the motor will guide

If the eye on the right sees
Then the left motor, in fact,
Will go ever fast
And the organism attacks

If eye and motor are on the same side
The cockroach always runs and hides
When eye and motor are oppositely attached
Humans, when stimulated, generally attack.

Cockroach research ideas
    Interestingly, if you look up topics such as research that needs done on the "choose-an-animal?", or "what do we not

know about "choose-an-animal?", you will find exclusively lists of things we already know about the animal you chose. It is impossible to list things we don't know. However, once you leave the common organisms you will find we know extraordinarily little about any animal.

It is impossible to make a list of things humans don't know, because we don't know them yet. That is why it is so frightening that so many of the highly educated people in Washington and in business think they know enough to run everything. Central planning will always fail because of this, but it can last a long time and cause a lot of misery and heartbreak before it collapses.

While there is a lot known about cockroaches already, most of what is known comes from information about the very few species that invade homes and cause concern. There is great variation in all species that are usually unknown. In fact, just searching for wild species and documenting what is found in the area is a great project.

Here are a few other ideas:

1. Survey for local species, map them and take notes on their habitat.
2. Collect local regional; species and make a display showing the variations.
3. Try to raise local non-pest species in captivity. This may sound simple and is for pest species. But very few animals can be, or have been, raised in captivity.
4. Raise cockroaches and observe behavioral characteristics.
    a. Can they complete a simple "Y" maze?
    b. Can they "remember"?
    c. Do they respond to sound, vibrations, light, color etc.
    d. Do they respond to a magnetic field?
    e. Do they respond to electrical fields?
    f. Food preferences?

    g. Compare all these results between immatures and adults or males and females.
    h. Behavior over twenty-four hours.
5. Almost all cockroaches I have studied have had internal parasite. If you have access to microscopes, it would be fascinating to find out which ones, how many etc. A few are intermediate hosts for other parasites of other animals.
6. Anatomy and physiology are well known for the pest species. But even then, new approaches can shed light on their biology. For example, determining pH of the gut at different times of the day, life cycle, or gender may all differ.
7. Determining the life cycle and seeing how it might vary under different conditions (temperature, light, pressure, etc.) would require much ingenuity.
8. Of course, determining what bacteria they may carry, analyzing food from fecal droppings, and investigating allergic responses are important possibilities.
9. It is easy to take blood samples from cockroaches and the study of insect blood is fascinating. In some ways similar to our blood but very different as their blood does not carry oxygen.
10. Common species of cockroaches can be obtained from pet stores or biological or scientific supply businesses. There are people who raise insect for fun (but I doubt profit). You might check out https://shop.bugsincyberspace.com/Cockroaches_c4.htm

CHAPTER 7 - COLEOPTERA

Phylum: Arthropoda
Class Insecta:
Order: Coleoptera:
Species: *Dendroctonus ponderosae*
Common name: mountain pine beetle

"A beetle may or may not be inferior to a man –
the matter awaits demonstration.
but if he were inferior by ten thousand fathoms,
the fact remains that there is probably a beetle
view of things of which a man is entirely ignorant."
G. K. Chesterton

Science Fair

They say diamonds are a girl's best friend. I gave my wife a beautiful beetle once. It was from Brazil, and I had embedded it in plastic. I think she would have preferred a diamond. She apparently didn't know, or maybe she didn't care, that beetles and diamonds have something in common. However, diamonds have nothing in common with plastic. Maybe that was the problem.

But ask Lauren Richey of Springville, Utah about beetles and diamonds. For her, beetles have turned into gold. Or am I mixing my metaphors here? It all started with that annual spring right called the Science Fair. Richey started doing science fairs in junior high. At first, she was not terribly successful, but then she began to win regional and national awards for research projects, usually involving light and photonics.

As a senior in High School, Richey read a paper suggesting that iridescent butterflies might contain photonic crystals. She admits that she read the article mostly because of the beautiful blue butterfly on the front. But that stirred her interest, and she approached John Gardner, a professor in the BYU Physics Department, with her idea to examine a beetle, *Lamprocyphus augustus*, a shimmery green beetle, with an electron microscope. And sure enough, the beetle exoskeleton contained a structure similar to photonic crystals.

Excuse me, but when I first heard this story, I didn't know what a photonic crystal was. Apparently, these structures, which are relatively rare, resemble the arrangement of carbon atoms in a diamond crystal. This crystalline shape can affect the propagation of electromagnetic waves in the same way that a semiconductor can in a computer. This would allow a computer to operate on light waves instead of electricity, a coveted goal in computer science. While the crystals in a beetle are too fragile for use in a computer, they might serve as a template for the manufacture of more sturdy structures.

But the significance of this story isn't about photonic crystals. It is about a student who did college level research while a senior in High School and published a scientific paper as a freshman in college. Since then, Lauren, a sophomore in physics at BYU has published two more research articles in professional journals and taken over 13,000 electron micrographs of beetles. She is presently examining beetles that contain photonic crystals of an opal like nature rather than those like a diamond.

The Intel International Science and Engineering Fair (Intel ISEF) is the world's largest international pre-college science competition. In Colorado, local and regional competitions are organized by the Colorado State Science Fair, Inc. (CSEF). Individual schools run a series of local competitions. Mesa State College hosts the regional fair for the western slope. These science fairs have been happening in Colorado since 1955, but CSEF was only incorporated in 1977.

The science fair may be one of the best kept secrets among middle school and high school students. Students can receive many different awards, in numerous categories, as well as get a taste of scientific research. Many students receive cash prizes or scholarships. In some cases, like Lauren's, it can set the stage for an entire career.

The regional Science Fair is generally held each spring, often at a local or regional college. And if you are looking for inexpensive fun, public viewing of projects is scheduled during some time of the weekend. Parent support, interested teachers, and mentors are essential parts of this activity. It is fascinating to see what young minds can do.

Anyway, I was sure to show my wife this story since I think she never properly appreciated the beetle I gave her. She's probably sorry now that she knows all about photonic crystals. She also never wore the mosquito ear rings, or the silver cockroach pendant I gave her. Wait. Silver cockroaches! That gives me an idea.

Beetles
The Coleoptera, with about 400,000 species, is the largest of all insect orders, constituting almost 40% of described insects and 25% of all known animal life-forms. They are generally a little larger than some insects and often brightly colored. Thus, they have been favorites of collectors and laypersons. Because of their abundance they are often readily available for science experimentation, and many can be raised in captivity.

They are fabulously variable and often colorful. If God doesn't love beetles, I don't know why he made so many. Oh, I know many people think they just sort of happened. But they really have very little in their makeup or biology to account for their variability or success at survival. I prefer to think that God is just a very dedicated "Coleopterist".

Consequently, they are represented in this book by three different species, more than most. The species I have included were basically just chosen because I have some small personal experience and knowledge involving them.

Blues Stain Pine
It's my wife's fault for having too many grandchildren. How was I to know that having children would end up being so complicated? I just wanted to have fun playing music with my grandkids, so I started building mountain dulcimers for them. Dulcimers are simple for children and adults to learn to play.

But the grandkids keep coming and it's all starting to get expensive. Do you know how much hard woods cost? Then I discovered a beautiful wood in Colorado that is less expensive and can be used to make dulcimers: blue-stain pine!

Nature often creates beauty through natural processes that seem ugly or harsh. Take a brightly colored mushroom growing amidst decaying organic matter, for example. Or a gold, metallic-colored beetle rolling a ball of dung. The geologic terrain is shaped through violent storms, earthquakes, and volcanoes. In the natural world, one lives when another dies. Vegetation is eaten to sustain life, and waste is produced to sustain vegetation.

In recent years, Colorado has experienced an infestation of *Dendroctonus ponderosae*, the mountain-pine beetle. Numerous trees in our pine forests have succumbed to this infestation. Their grey shapes can often be seen looking like ghosts in the dense green foliage of the forests.

The adult, female beetles chew through the bark of available trees while releasing pheromones, a gaseous attractant for male beetles. They mate and lay eggs within tunnels chewed into the trees. The eggs hatch, and the larvae continue to eat and burrow through the trees. Eventually the larvae pupate and hatch out the next year as new adults.

Once trees are infected, pheromones attract more beetles to the same trees resulting in a mass infestation that can eventually kill them. One cause of trees death is the numerous galleries and tunnels through the trees that inhibit water and nutrient flow.

Trees have some natural abilities to combat such infestations. One defense mechanism is the resin that's secreted by pine trees, commonly at the site of wounds. However, pine beetles carry, within a pouch in their mouthparts, a variety of Ascomycete fungi that can interfere with the tree's resin production.

The fungus produces a thread-like mass of "hyphae" and spores that are "sticky". These eventually block the water-conducting columns of the tree, drain the trees of their nutrients, and eventually cause the tree to starve to death.

Beetle larvae feed on the fungus, as well as the tree, further promoting beetle growth and increasing their numbers.

The fungal hyphae are made of inert material that persists after the fungus dies. These colored hyphae cause discoloring of the wood. While the wood is often discolored bluish, different fungi can stain the wood other colors, including shades of blue, brown, green, red, and even black.

If a tree killed by pine beetles can be harvested quickly though, the wood can still be used. However, with too much time lapse the wood begins to dry, crack, and warp due to the many tunnels left by the beetle. These problems make it difficult to use the lumber for building. Many of the infected trees infected are older, with diameters greater than eight inches. They are often scattered within the forest making it difficult to harvest them.

The death of an ancient tree is sad, but natural. The association of three living organisms that complete their life cycles in tandem is an arrangement called symbiosis, living together. Living things tend to live on or in other living things, and such associations as these may be the most common form of life on earth.

What is the result? A beautiful, blue-stained wood called blue stain pine. Have you seen it? You may have seen it as rustic siding and paneling on cabins or homes, or in mountain dulcimers made in Colorado. I find it fascinating the way nature creates beauty through natural processes that sometimes seem ugly and harsh.

Shields and Heraldry

One of the distinctive features of human history has been the shield. As men developed the inability to get along with each other and a tendency to fight and kill off their own species, they began developing weapons. Shortly after that they began developing protective gear. Undoubtedly a form of shield would have been the easiest and one of the first such implements.

As practical and useful a shield must have been, it probably was not long before someone decided to decorate the shield in some way. Perhaps the first impetus was to strike fear into an enemy. Or maybe it was to identify themselves as a friend or foe. Or, knowing the human desire for recognition and attention, perhaps they earliest shields were simply a form of brand identity and individual expression.

The distinctive shield shape has been used in heraldry for centuries, perhaps since ancient days. Further, as a front for individuals and armies it invited symbolism and identity for the individual or armed colleagues. But colorful and symbolic designs are nearly a unique expression of humanity.

I say nearly because colorful advertising and symbolic representation is common in the biological world. Flowers advertise their nectar in color and many animal use decorations either to hide or to warn other animals away.

But only one animal uses a shield. The Coleoptera.

Beetles in general

Like most insects, beetles have an external skeleton, but theirs is particularly hard. They have two pair of wings, also like many other insects. However, the first pair of wings in beetles are modified into a hard shield-shaped structure that covers the second pair of wings and much of the body. This structure is called an elytron. The term is derived from a Greek word which means sheath. The plural is elytra. While early scientists called it a sheath, we know they meant shield.

Beetles can fly but most are awkward and don't fly far or often. Their hardened exoskeleton provides them with some extra protection. Many of them have also developed either foul smelling odors or poisonous compounds that predators avoid. Those that do this often have distinctive markings so that predators learn to recognize them and leave them alone. This has led to other, nonpoisonous species developing similar markings, so they benefit from predator avoidance. Other beetles have also adopted camouflage colorations that help them hide in the habitat.

All in all, their common distinctive colorations benefit them in a number of ways, in much the same way that heraldry has

> **While early scientists called it a sheath, we know they meant shield.**

benefitted mankind. Such shield coloration announces kinship, relationships, warnings, and protection. A curious development in an animal of small size and very different biology, anatomy, and physiology than humans.

SEE HERE MY SHATTERED SHIELD

I cannot speak yet have so much to say.
I cough and choke on lungs of clay.
I have seen all the life that I have lived.
I've given all that I can give.

See here my broken blade,
Used to shear her auburn braid
That she might not be betrayed.

Four things change like the seasons,
Time and knowledge, rhyme, and reason.
You are young and think you know
How the way of your world will go.

See here my dented helm,
Used to protect my queen and realm,
Until I was overwhelmed.

I was hers, and she was mine.
Children arrived over time.
We claimed the woods and the land,
And all prospered from our hands.

See here my shattered shield
With blood and mud now congealed.
She escaped because I did not yield.

EVERYTHING THAT CREEPETH ON THE EARTH

Now at las I can hide my face.
This cloak of darkness I embrace.
I have withstood the enemies blows.
I have conquered demon foes.

See there's the gauntlet cold.
It served me well, true, and bold.
Now it and I turn to mold.

Though fleeting moments, I have endured,
Of this much I am sure,
My heart swears it's not been wrong.
I long to see you, it's been so long.

See the lance shattered in two
In spite of all that I could do.
Each, in the end, receives his due.

Hand in hand we stood out ground
Surrounded by baying hounds.
Even yet, with passion spent,
Until our very lives were rent.

See the offering that I bring.
A new world born in suffering.
My pledge to thee is all I sing.

I once was trapped in life's parade.
It all seems now a promenade.
Seeming a King without a crown.
In retrospect I seem a clown.

I have stood where no man goes.
I now know what no man knows.
Ans I must go where all men go.

See here my shattered shield
With blood and mud now congealed.
Unto my God and Christ, I now yield.

LADY BUG

Family: Coccinellidae
Class: Insecta:
Order: Coleoptera:
Species: *Coccinella septempunctata*
Common name: Lady Bugs

Lady beetles
According to legend, the common name for these beetles refers to the Blessed Lady, the Virgin Mary. At one time during the Middle Ages European crops were being destroyed by pests, probably aphids. The famers began praying to The Virgin Mary, the Blessed Lady. Eventually the farmers began to note red and black beetles in their fields and the crops were saved. They purportedly named the beetles "our lady birds" or lady beetles. In Germany these beetles are sometimes called Marienkafer, Mary's Beetles.

While not all lady beetles are red in color, many of them are and they usually support several black dots. The seven-spotted lady beetle is believed to be the one named for the Virgin Mary. The red color is her cloak, and the seven dots represent her seven sorrows taken from Catholic theology.

The reason Lady Bugs can sometimes be beneficial is because they have such voracious appetites. A ladybug may consume 5,000 aphids in its life. As soon as ladybugs hatch, they begin to feast. If you're unfamiliar with ladybug larvae, you would probably never guess that these odd creatures are young ladybugs. They look a lot like miniature alligators. They have long, pointed abdomens, spiny bodies, and legs that protrude from their sides. The larvae feed and grow for about a month, and during this stage they can consume up to four hundred aphids even before they are adult.

But Lady Bugs don't just eat Aphids. They also eat fruit flies, thrips, mites, and other plant-damaging insects. However, different species of Lady Bugs prefer different foods. While many prey on garden pests, the Lady Bug called the Mexican Bean Beetle, and another called the squash beetle, also feed on the plant leaves mentioned in their names. Those species may not be as welcome.

They are also extremely prolific, and one ladybug lays literally hundreds of eggs. However, because they are carnivores there can never be as many of them as there are prey organisms. It takes more than one zebra to grow a lion. However, once they are present their numbers can increase amazingly fast.

While ladybugs are common and seem gentle, if they feel threatened or startled, they can emit a foul-smelling liquid from the leg joints. This is formed from their blood which is called hemolymph. This can also leave a yellow stain on surfaces. They can also ooze alkaloids from their abdomen. This mix of secretions makes them unpalatable to potential predators.

So, like many other insects, ladybugs use distinctive coloration to signal their toxicity to would-be predators. This may explain why their typically brightly colored and distinct patterns exist. Insect-eating birds and other animals learn to avoid meals that come in red and black and are more likely to steer clear of a ladybug lunch.

LADY BUG

Oh, to fly on hidden wings
With seven spots and be the queen
Of all the garden under the sun
And then fly all home when day is done

With scarlet cloak befitting queens
And seven spots upon her wings
For the seven joys that she knows
And for her experience of seven woes

The gentle touch and fearless mien
Then she hurries home again
Moved upon by an ancient choir
Telling her that her house is on fire.

Old Testament
Leviticus 11:22
Even these of them ye may eat; the locust after his kind, and the bald locust after his kind, and the beetle after his kind, and the grasshopper after his kind.

FIREFLY

Phylum Arthropoda
Class Insecta:
Order Coleoptera:
Family Lampyridae: - lightening bugs, fireflies

<u>Truth in advertising</u>

Maybe I should mention this first. Fireflies aren't flies. They are a part of the beetle family. They are part of a mostly nocturnal group of beetles called the *Lampyridae* (Greek for "to shine"). However, fire beetle doesn't sound so romantic, does it? There are more than 2,000 species in this group and only some of them can shine like a lantern.

> **Fireflies aren't flies.**

One more thing. They taste disgusting and should not be eaten. If you do try to eat a firefly, it will probably taste very bitter as, when attacked they shed drops of blood that has chemicals that make the bitter and slightly poisonous. Most animals have learned this and avoid munching on fireflies. Which may be another advantage of lighting up in the dark which you think would make them attractive prey.

Light

Most of the time we think of light as producing heat. But fireflies are bioluminescent and produce light through a chemical reaction that produces extraordinarily little heat. In fact, producing a flash of light increases their metabolic rate by as little as thirty-seven percent. They produce light for a variety of purposes: signaling each other, warning predators, finding mates, and attracting prey.

Fireflies' lights are the **most efficient lights in the world.** One hundred percent of the energy created is emitted through the light. In comparison, an incandescent bulb emits 10 percent of its energy as light and a fluorescent bulb emits 90 percent of its energy through light. Fireflies' efficiency is partly due to luciferin's heat resistant properties. Luciferin is the enzyme they use to create their flashing light.

Their light is created by a chemical reaction during which oxygen combines with calcium, adenosine triphosphate (ATP) and luciferin with the help of the enzyme luciferase. When luciferase was first discovered, the only way to obtain the chemical was to extract it from fireflies themselves. However, scientists have since discovered how to make the enzyme synthetically.

Luciferase has been used scientifically as a marker to detect blood clots, to tag tuberculosis virus cells, and to monitor hydrogen peroxide levels in living organisms. It is also used in scientific research, for food safety testing, some forensic tests, and in genetic recombination experiments.

Each species has their **own specific color** of light they produce. Some glow blue or green while others glow orange or yellow. Most of the fireflies flying around are males looking for a mate. Each species has a specific light pattern that they use to communicate with each other. Once a female spots a male she likes, she will respond with the same light pattern. Usually females are perched on plants, waiting for a mate. When they are larvae, fireflies **use their** bioluminescence to scare off predators.

Scientists aren't sure why some fireflies synchronize their flashing. Not all do this, but some do. The only species of fireflies in America that do this are the *Photinus carolinus* that live in the Great Smoky Mountains. The U.S. National Park Service sometimes organizes **watch parties** for these shows.

Not all fireflies have the "fire". Fireflies in the western states don't light up like those in the east. Those who don't produce light are usually most active **during the day**. Non-bioluminescent fireflies use pheromones to attract mates.

The following are two contrasting poems about fireflies using similar meter and rhyme schemes.

CHEMILUMINESCENCE
(a poem by Gary McCallister)

They make their light by chemiluminescence
That alone explains their flashing brilliance
And also explains their fleeting transience
They twinkle somewhat like a star
A flash of light like pulsing radar
But that's no longer where they are

FIREFLIES IN THE GARDEN
(a poem by Robert Frost)

Here come real stars to fill the upper skies
And here on earth come emulating flies
Although they never equal stars in size
(And they were never really stars at heart)

> Achieve at times a very star-like start
> Only, of course, they can't sustain the part

Frozen Light

Don't ever ask your children "why" they did something. As a matter of fact, don't ask politicians, criminals, or your spouse "why" they do the things they do. There are only two possible responses to this inquiry. They will either stone wall by shrugging their shoulders and muttering, "I dunno" or they will redirect your attention to someone else like this, "He hit me first."

Stonewalling can sometimes be overcome by intense and prolonged pressure. However, it pits your desperation to understand against their desperation to not be caught. You will lose.

The blame response is even more difficult because it invites you to ask why the other sibling hit first. That answer will either be "I dunno" or another excuse directing your attention to yet another event. I know from experience you will eventually forget what the whole interrogation was about!

If you enjoy this kind of interaction, you might have the makings of a scientist. If you decide you want to know more about a subject, you will either discover that no one knows or cares, or that there is something you have to understand first before you can understand what you wanted to understand.

So, following a recent electric storm, I wondered how much light was in a lightning flash. I couldn't find an answer in my biology books although I did find out that a firefly puts out about one fortieth of a candle worth of light. Candles vary quite a bit, so I looked up beeswax candles specifically. They produce about 13 lumens, so fireflies emit 0.325 lumens, or one fortieth of 13, of light.

However, fireflies don't emit a constant light, so you couldn't really read a book with forty fireflies. I suppose you would need hundreds, if not thousands, to be useful. Lumens are a measure of the "amount" of visible light that comes from a source, so that depends on how close to the fireflies you are.

The actual quantity of light is a little hard to determine because a piece of light is so small. I think it may also be because a piece of light travels at 186,000 miles per second, so you had better be quick about it.

That is why physicists invented the "quantum" to measure such things. A quantum is the minimum amount of any physical entity involved in an interaction. It's like a penny would be the quantum of the financial world. Anyway, a quantum of light is called a photon. (If you have more than one quantum of anything, it is called a quanta. Personally, I just call a quanta of light "brighter".)

Then I discovered something really amazing! Dr. Lene Hau of Harvard University has been able to slow a beam of light down to 17 meters per second. Good grief, what does slow light look like? Even more amazing, she later stopped a beam of light dead in its tracks! She says that when you stop a photon completely, it turns into matter. Then if you let it go it turns into light again. ("Toto, I've a feeling we're not in Kansas anymore.")

I have some questions. What does a piece of frozen light look like? Can I freeze light so I can read by it later? Wait! If I freeze some light now, can I use it next January when it's dark outside? Now that would be saving daylight. Would thawed light change the direction of my shadow since it came from a different direction relative to where I am standing at the moment? Would light frozen in the summer cause plants to get confused if it was thawed in the winter? What if I freeze light in the northern hemisphere and thaw it in the southern?

Don't ask me why I am writing about fireflies, quanta, and frozen light. I was just trying to find out how much light is in a lightning flash.

Old Testament
1 Kings 7:4
And there were windows in three rows, and light was against light in three ranks.

THE SCARAB OF RA

Phylum: Arthropoda
Class: Insecta:
Order: Coleoptera
Family: Scarabaeidae
Species name: *Scarabaeus sacer*
Common name: the sacred dung beetle of ancient Egypt

"If one could conclude as to the nature of the creator from a study of creation, it would appear that God has an inordinate fondness for stars and beetles."
John B. S. Haldane

Delicate subjects

I hope I can handle this delicate topic tastefully. My scientific education included specialization in the study of parasites; intestinal parasites, to be specific. Such specialization places me at the absolute nadir of science, below which it is impossible to go lower.

However, it has also made me somewhat of a specialist on biological waste. My knowledge has come in handy in growing my garden as I can usually find the highest-quality steer manure for fertilization. "Steer manure" seems a preferable term to the more common designation of this material for use in a family paper. But it would be very difficult to talk about Scarabs without mention dung, cow chips, horse apples, manure and other euphemisms for animal waste.

The Scarab beetle was apparently worshiped in ancient Egypt. If not, it was certainly a popular ornament for clothing, jewelry, and on buildings. However, the Scarab is a dung beetle that feeds on dung for much of its life. In fact, a dung beetle can bury dung 250 times heavier than itself in one night.

What happens is that they roll the dung into a ball and roll it over to, and insert it into, a hole in the ground they have dug. They then lay their eggs on it and the larvae feed on the dung. This leads to occasional observations of dung beetles rolling dung balls and, later, numerous baby beetles scurrying out of the ground all at once.

Presumably, Egyptians thought this was akin to the sun that came out of the ground every morning and rolled across the sky until night until it was buried in the earth again. The sun was seen as the source of life and so they understood the dung beetle to represent the sun, or life itself.

Recent experiments have shown that some dung beetles navigate rolling their balls of dung back to their home using the stars of the milky way. Dung beetles are less prevalent in cities and now it isn't clear if that is the case because of a lack of dung, or light pollution. Maybe it's both.

> **Presumably, Egyptians thought this was akin to the sun that came out of the ground every morning and rolled across the sky until night until it was buried in the earth again. Was this a true belief or a metaphor?**

THE SCARAB OF RA
What is this? Tiny beetles
Hurrying from a hole in the ground
One after another they clamor out
A cacophony without sound

What is this? The flaming ball
Rolls steadily and reliably across the sky
Yet there are no wheels or gilded chariots
No visible wings with which to fly

What is this? The golden scarab
Providing food and shelter for its young
But the golden ball and symbol of life
The claims of Ra rest upon dung

<u>Research Projects for Coleoptera</u>

I wish I could make a list of the things we don't know about beetles. However, I don't know them yet so I can't. Besides it would be extremely long. We do know great deal about beetles because they are numerous, colorful, sometimes important economically, and many are model organisms.

Model organisms are living organisms that are easier or safer to maintain or handle on which scientific experiments can be conducted. There are frequently used for medical, agricultural, or environmental studies. Many insects serve this purpose, but perhaps beetles are more common than many others.

The coleoptera have a worldwide distribution, environmental significance, the possibility of extrapolating research studies to vertebrates, and the relatively low cost of rearing. Two wonderfully distinctive advantages are that no one usually cares what you do to them, and failed experiments can just be flushed.

The species listed above may not be the best organisms to attempt to do research on. Your choice of research projects may revolve around the species you want to know more about, or what is most available to you. Various species of Tenebrionidae such as *Tenebrio molitor* (the darkling beetle, or *Tribolium confusum* (the confused flour beetle), both of which are common grain pests and easy to raise.

Much the same kinds of studies that were done on cockroaches can also be done with beetles to good experience and knowledge, especially if you are dealing with species that are not economically or medically important. Extraordinarily little is known about free-living beetles. Even learning how to culture them and keep them alive in captivity is useful information for someone.

Some easily approached projects with non-pest species might be the following:
1. Collections are always valuable.
2. Interview local agriculture leaders to learn about local beetle pests.
3. Compare densities of beetle populations in various environments.
4. Some beetles produce volatile compounds. Do they all? Which ones?
5. Do beetles respond to volatile compounds? Do they use this ability to find plants or food? Are they repelled by some compounds?
6. Most beetles are soil or liter dwellers. Is there sound in the ground? Can sound be used to track them, find them, or repel them?
7. What effect do common pesticides have on non pest populations and species of beetles.
8. What effect do non-pest beetles have on local soil?
9. What effect do non-pest beetles have on plants.
10. Many beetles have nematodes and other organisms in the digestive system to help them digest their complex carbohydrate diets. Do they all? Which ones?
11. Variances in blood cells between types of beetles.
    In addition, here are a few suggestions about specific beetles addressed in this book.

Lady Bugs research ideas

The number of lady bugs have declined for many years and there is a great need to find those that are around, identify them, determine their locations and numbers. Collections reduce the population so in is better to photograph them and cooperate with something like the Cornell University Lady Bug project. Much helpful information can be found on their site. http://www.lostladybug.org/participate.php

Some lady bugs can be reared in captivity for helping the environment. And no domestic species from different ecological areas are of great interest and may be declining without general knowledge.

Fireflies research ideas

If I were going to do research on fireflies I would stay away from the subject of light. See, the big boys are all over that with genetic engineering and biochemical reactions. You'll spend years reading the past literature before you can even start. And if discover anything you'll just be an *et al* on a paper with seventeen authors. Maybe that's just me. Of course, I live out west and most of the fireflies where I am don't have any fire anyway.

I'd probably start with just trying to find some. That alone won't be easy since they don't go around flashing gang signs and stuff. But just finding the ones in your neck of the woods might be interesting. There may even be some around no one knows about. And the ones that are there may not have much information about them. People tend to do research on things they think are important and they seldom think invisibles things are important unless someone shows them that they are attacking them in some way.

I find it interesting that humans are seldom as interested in things, invisible or otherwise, that are quietly beneficial to them. Unless it makes them money, they tend to not see any value, even though it may be essential to the wellbeing of the earth.

Anyway, here are some things that might be worth knowing about local fireflies.
1. Are there any in your area?
2. How many different kinds?
3. How can you tell them apart?
4. Where and how can you find them?
5. What do the eat?

6. Where do the live?
7. What environment do they prefer?
8. What is their life cycle?
9. What time of years can they be found?
10. Is it different for different species?
11. What do they eat? What is the difference between males and females?
12. What do the immatures look like?
13. Where do the mate and lay eggs?
14. Are they harmful to crops, animals, or the environment?
14. In what ways do the benefit the environment?
15. What temperature extremes can they withstand?
16. For how long?
17. Can they be raised in captivity?
18. Do they have internal or external parasites?

CHAPTER 8 - HYMENOPTERA

Phylum Arthropoda
Class Insecta:
Order Hymenoptera:
Family Apidae:
Apis mellifera - Honeybees

It's not how busy you are, but why you are busy.
The bee is praised.
The mosquito swatted.

There are two entries for the Hymenoptera. That is because there is more than one kind of bee. And the difference is more than just details of anatomy. Both kinds of bee live entirely different lives and serve entirely different purposes.

<u>The worth of insects</u>
Some people say that this is the "age of man", but that is probably just because men get to say such things. If insects got to write books, they would probably say this is the "age of insects." Estimates from various sources differ, but most experts think the number of insect species is close to a million. However, there are at least twice that many that have not been identified. Some people think there could be as many as thirty million different species. They probably represent about eighty percent of the known animals in the world. At any given time, there are 10,000,000,000,000,000,000 individual insects alive (that is ten quintillion).

I don't know if mere numbers make insects any more valuable than men, but from a biological perspective their numbers are impressive. They must be doing something right, and they have been doing it for a lot longer than humans. The earliest insects seem to have originated about four hundred million years ago. The earliest human-like critter is estimated to have lived around two hundred thousand years ago. Now I am the first to admit that age doesn't always produce wisdom (just ask my children). However, insects seem to be good at staying alive.

I don't know whether this is the age of man or the age of insects. I am not even totally clear on what age I am. When I recently asked my doctor why I was having trouble with my eyes, I was told that I was old. Then he charged me money for that. I told my wife that I was a child of God. She agreed but added that I was about equivalent to a two-year-old.

So being old, and therefore wise, I still think deciding which animal is more important than another is sort of beyond me. We are all so tangled up together here on the earth. If we want to talk about the value of one animal over another, though, I have a suggestion.

According to the USDA, the value of pollination to agriculture in the United States was fifteen billion dollars in the year 2000. Then, according to the American Bee Journal, there were approximately two million five hundred thousand beehives in the United States, in 2010. While these data are separated by ten years, they are the best I could find. But that puts the value of a single beehive at close to $600.00 per hive, and that is just counting the hives value to the public in pollination.

According to Bee Culture magazine the value of honey produced in the United States last year was close to three hundred million dollars. That, divided by the number of hives, adds another $120.00 to the value of the hive for a total of $720.00 per hive. Of course, there is also a market for bees, bees wax, pollen and propolis. Now most beekeepers can only dream of making $700.00 per hive. Like the old saying goes, "Beekeeping is a good way to make a million dollars. Just start with three million dollars."

Then I attended a seminar a couple of weeks ago. The presenter said that Pennsylvania State University had conducted a study and found that one beehive was worth about $13,000.00 dollars to the public in goods and pollination. (I have to assume their data is better than mine. After all they are a university.) Of course, the study was measuring the entire benefit to the public. Since bees can forage for up to five miles, and routinely forage for three miles, many gardens and commercial famers benefit from pollination they never pay for. That seems to be the big difference between our estimates.

I don't know if that makes bees more important than any other animal. To humans we measure significance by GDP whereas a single flower pollination is a question of survival. It does sort of indicate that maybe communities and neighborhoods ought to be encouraging beekeeping, not discouraging it. Just for the public good.

> Most beekeepers can only dream of making $700.00 per hive. Like the old saying goes, "Beekeeping is a good way to make a million dollars. Just start with three million dollars."

Keeping honeybees

I have kept honeybees three different times in my life. You know the old adage, "If at first you don't succeed, try, try again."? Well, if you don't succeed after the third time you should probably give up. That's what I have done.

My first-time keeping bees I was just a kid, and the truth is it was my grandfather that was the beekeeper. I just went along and helped him until he got tired of my help and sent me home. He's the one that taught me to talk to the bees. One day he told me he could talk to the bees and bent down and said in a real breathy voice, "Hello, Bees." The hive buzzed back at him. I asked what they said, and he said they said hello. Then he said, in the same breathy voice, "Bees, I want you to meet my grandson, Gary." The hive buzzed. He said,

"They said how do you do." I think I mumbled, "Hi", or something. What does one say when introduced to bees?

I hate to say this, but sometimes you couldn't always believe my grandfather. My Mom told me that he was a rascal when she was growing up, but he was always kind to me. Anyway, later that day I went over to the beehive to see if I could talk to the bees. I bent down right in front of the hive and in my best breathy voice said, "Hello, Bees." One bee immediately came out and stung me right on the lip.

Much later in life, when I was married and with children, I started a couple of beehives to teach my children how to talk to the bees. They say abuse is intergenerational. The problem with beekeeping, colored by fond memories and images of country leisure, takes a little more time and energy than you might expect. And like with keeping a lot of animals, the time and energy required happen on the timing and needs of the bees, not the beekeeper. I was teaching at the university as well as running a mosquito control district and the bees needs just didn't come first so the bees gave up on me.

The third time I kept bees started just a couple of years before my retirement. I decided that keeping bees would be a good part-time job for when I retired. Besides, by then I had some grandchildren and I wanted them to learn how to talk to the bees. That time I stayed with it about ten years and eventually had about twenty hives and four grandchildren helping me, until I got tired of their help. It worked out pretty well except it turned out to be more like a good full-time job. Also, it became obvious that beekeeping probably wasn't a career any of them wanted to follow for an occupation in this modern world.

As they made better career decisions than I did, they left for university and jobs and suddenly I had twenty hives all too myself. I discussed it with the bees, and we came to a mutual decision to sell the company to the employees and let them all swarm and go wild.

Hard work?

Honeybees are frequently used as examples of hard-working creatures. The truth is, compared to many other insects, they are kind of lazy. On the other hand, compared to many other insects, and humans, they are rather industrious. I think it is more appropriate to think of beekeepers as examples of industry. You probably think you spend a few hours setting up a bee yard and hive and then go sit on the porch until fall when you go spend a few hours collecting honey.

Let me tell you a little about beekeeping as an activity. Every beekeeper knows that on stormy days, whether cloudy, windy, or rainy, honeybees tend to stay home. Oh, a few might venture out here and there. But you don't want to open a hive on these kinds of days. There is a lot more bees than usual, and they are cranky. So, you can't work on blustery, windy, rainy, or cold days.

In fact, the best time to work with the beehive is after ten o'clock in the morning and before about four in the afternoon. They seldom really get moving and busy until ten. Between ten and four the greatest number of bees will be gone from the hive, finally doing their work, and the remainder will be busy keeping up with the foragers. They are far more willing to ignore a pesky beekeeper poking around in the hive during the hot, middle of the day.

The also tend to knock off early. After about four or five o'clock in the afternoon, the number of foragers decrease and the number of bees in the hive increase. This means that beekeepers' busiest part of the day is also the hottest part of the day. And it isn't as if you can wear shorts and a T-shirt for the job either. The suit of armor required for doing the job, while made of canvas, is still very heavy and inhibits air movement.

In addition, a full honey super can way upwards to sixty pounds. Dressed in full body armor, lifting heavy weights in the heat of the day, while being pestered by flying insects who can sting if they smell fear, makes beekeeping a difficult, although critical, job.

> **Old Testament**
> **Judges 14:8**
> ". . . he turned aside to see the carcass of the lion: and, behold, there was a swarm of bees and honey in the carcass of the lion."

The organism

There is a lot more than one kind of bee in the world. Most people are familiar with honeybees, and while honeybees are similar, there are also several varieties of honeybees.

The really interesting fact about honeybees is that the individual bee is not the organism. The hive is the organism. If one captures a foraging bee and puts it into a jar with water pollen and honey, it will be dead within a few hours. Bees cannot exist outside the hive.

It's as if the organisms were this collection of cells organized into a series of individual bees that had to live together, but that fly apart on demand to take care of bodily functions and then reassemble.

I have toyed with the idea of creating an adventure comic series called Beeman. If the bad guys try to kill him, he would, simply fly apart into 50,000 little pieces, sting the heck out the bad guys and then reassemble. However, I am way to busy writing these serious, informative books about science, so I am giving anyone who is interested the idea for free. That's just the kind of guy I am.

Honeybees are colonial insects that live in a colony called a hive. There is one queen, and series a of female workers and some male drones. They are known for construction of perennial, colonial nests from wax, for the large size of their colonies, and for their surplus production and storage of honey. Presently, only seven species of honeybee are recognized, with a total of 44 subspecies. The best-known honeybee is the western honeybee which has been domesticated for honey production and crop pollination.

Humans value bees wax for candle making, soapmaking, lip balms, and other crafts. However, the wax is also useful in many other craft applications to one degree or another. Some other types of related bees produce and store honey and have been kept by humans for that purpose, including the stingless honeybees, but only members of the genus Apis are true honeybees. The study of bees, which includes the study of honeybees, is known as melittology. But honeybees represent only a small fraction of the roughly 20,000 known species of bees.

As an exploration of the concept discussed earlier about how insect enrich the world in many ways the following illustration, and paragraphs, attempts to generate some idea of this contribution using honeybees for an example. I have expanded on this topic here because honey bees are better understood than most insects and data is available mto make some reasonable estimates.

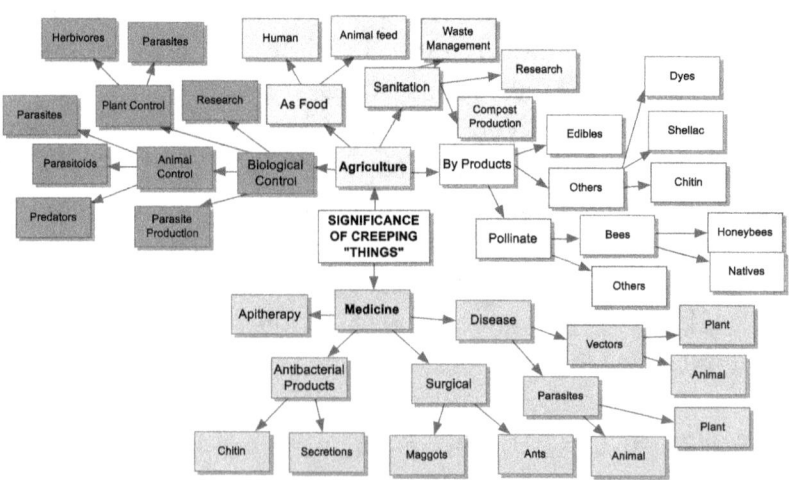

Honeybees as natural enrichment

The above discussion and estimates are to show the possibile and potential impact of insects on plant nutrition through rapid production of biomass, production of animal waste, and as a source of concentrated plant nutrients. Honeybees are used here simply as an example because they are well researched, and much is known about their life cycle.

Obviously, honeybees are not typical insects. They are larger than many, perhaps most insects, and may have unique changes in numbers, biomass, chitin production, and waste production over the course of time due to their unique life cycle. There are many insects that may outproduce honeybees in numbers, uniform growth rates, and biomass. In addition, many insects are decomposers and scavengers and thereby grow on what is normally waste products of the environment.

While honeybees are larger than many, perhaps even most, insects, larger numbers of individual smaller insects undoubtedly create an equivalent biomass. An insect of half the size but twice the numbers might be assumed to have an equal amount of organic waste. There are insects that produce even more generations per year than honeybees. In fact, if conditions are favorable, many insects, such as mosquitoes, can produce a new generation every two weeks. These are not even the most rapidly developing insects known.

Honeybee biomass

The number of honeybees in a hive can vary greatly. However, a healthy hive usually has between 20,000 and 50,000 during the summer. Researchers often estimate hive numbers as high as 80,000. I am going to use the lesser range for my estimates to accommodate for the far lesser numbers during the winter months. The midpoint of the lower range is about 35,000 so I am going to use that as an estimate for the number of bees in a hive, recognizing that the numbers go up and down in an annual cycle.

The annual reproduction cycle for bees also varies with the time of year. However, researchers suggest that the average bee lives about seven weeks. If we use this estimate then bees produce about seven generations per year, again depending on seasonal variation. These estimates suggest than one beehive produces about 7 x 35,000 bees annually, or 245,000 bees.

A single honeybee weighs about 0.00025 pounds (or 0.113398 gm). So, the annual production of honeybees, by weight, from one hive would be 0.00025 x 245,000, or 61 pounds of bees per year.

These bees all produce frass, die, and decompose, thereby enriching the environment with organic matter. The protein content of insect's ranges between 50 to 70 percent. Sugar content is as less than 10% and the many carbohydrates are complex and varied. Insects are rich in N, Ca, P, Fe, and Zinc. Insects also possess a variety of fatty acids, often as much as 50%, but are extremely low in cholesterol. Obviously decomposing bees represent a significant amount of environmental nutrients that are recycled.

Honeybee Fras (manure)

A previous study indicated that 5000 bees produced about 16 g of frass (insect waste) per m^2 per month. This would be 16 x 245,000/5000 or 49 g/m^2/month, or 588g/m^2/year. This also equivalent to about 0.0032 g/m^2/mo. or 46 g/m^2/year. This is about 0.1 lb.

Bee frass contains about 99% total nitrogen in organic form. The previous study on bee frass concluded that a colony of 20,000 bees foraging in a small area of about 3 m^2 provided about 70 g inorganic N/m^2/month to the environment. This would be about 84,000 g N/m^2/year. In pounds that would be 185 lbs.

Twenty-thousand bees is not a full healthy hive. But using the figures from above for the weight of a bee, 20,000 bees would weigh about five pounds. Since a normal hive will produce 61 pounds of bees, these frass estimates would need to be corrected for total weight, or 61 lbs./5 lbs. = 12. One healthy beehive would then contribute 12 x 84,000 g N/m^2/yr., or 1,008,000 g N/m²/yr. (2222 lbs.). This does not include the nutrient value of the bees' body.

Honeybee nutrient value

It is difficult to calculate the amount of nutrition and energy available in a single insect body due to variability of methods and the kinds of information desired. For example, how much total energy, or how much of a specific component such as nitrogen or a specific vitamin. However, there has been an increased interest in producing insects for human or animal consumption due to increasing pressure on the agriculture system worldwide.

The following table shows some of the results for various species that have been studied, including bees. The average energy per bee or wasp appears to be about 525 kcal/100g (525 kcal/0.22 lbs.). This would be 27,669 per hive per year would produce app 14,000,000 kcal/yr./hive. (See the table on the following page.)

Honeybee Chitin

Chitin is found throughout the exoskeletons of most insects, where it may be present in amounts ranging up to 60% in special parts such as the flexible portions. The average chitin content in the cuticle of a number of different species is reported to be 33%.

If one third of each insect's exoskeleton is composed of chitin, which presumably breaks down into some form of chitinase and other products that inhibit other insects and stimulates plants to increase defensive actions then the overall death of insects in the environment would seem to have some beneficial effects that have so far been undocumented.

If we assume that honeybees are typical in their biomass and frass production and apply those numbers to the number of insects as a whole, we get staggering numbers. This calculates 1.6×10^{16} g/mo. of frass worldwide. Decaying organic matter from dead insects, with accompanying chitin, would be 6.1×10^{31} pounds of biomass/year.

The contribution they make in total to the replenishment of the earth through waste material and decomposing organic matter, cannot be estimated. This also does not consider their roles in decomposition, pollination, as predators on enemy species, and in other yet unidentified ways.

Nutrients and energy*	Cockroaches (Blattodea)	Beetles (Coleoptera)	Flies (Diptera)	Beetles (Hemiptera)	Bees, wasps, ants (Hymenoptera)
Data amount n	3	45	6	27	45
Protein, %	57.30	40.69	49.48	48.33	46.47
min	43.90	8.85	35.87	27.00	4.90
max	65.60	71.10	63.99	72.00	66.00
SD	11.71	15.61	13.12	15.09	15.19
Fat, %	29.90	33.40	22.75	30.26	25.09
min	27.30	0.66	11.89	4.00	5.80
max	34.20	69.78	35.87	57.30	62.00
SD	3.75	18.91	9.35	18.74	11.96
Fiber, %	5.31	10.74	13.56	12.40	5.71
min	3.00	1.40	9.75	2.00	0.86
max	8.44	25.14	16.20	23.00	29.13
SD	2.81	6.50	2.81	5.74	6.32
NFE, %	4.53	13.20	6.01	6.08	20.25
min	0.78	0.01	1.25	0.01	0.00
max	10.09	48.60	8.21	18.07	77.73
SD	4.91	12.33	3.25	5.93	20.56
Ash, %	2.94	5.07	10.31	5.03	3.51
min	2.48	0.62	5.16	1.00	0.71
max	3.33	24.10	25.95	21.00	9.31
SD	0.43	4.83	8.14	5.44	1.56
Energy, Kcal/100g		490.30	409.78	478.99	484.45
min		282.32	216.94	328.99	391.00
max		652.30	552.40	622.00	655.00
n (Energy)	0	17	3	18	28
> 400 kcal/100g		13	2	13	27
> 500 kcal/100g		10	1	8	7
SD		111.42	173.28	98.53	58.88

HONEYBEES ARE LAZY

Honeybees are lazy when the morning is hazy
They might start to begin when the clock strikes ten
Often just one or two at the door at first
Will venture out to slack their thirst

But the day begins to hum a little past one
When bees begin to dangle between the flower's tangles
And the violins shimmer under summer sun
Until just before sundown when a bee's days done.

NATIVE BEES

Phylum Arthropoda
Class Insecta:
Order Hymenoptera:
Family:
Native bees, many species

Bottlenecks

My wife has house plants. When she waters them, she sometimes walks behind me, dribbles a little water on my head, and says, "Grow little flower grow". But let me squirt her with water from the hose just once and she won't help with the yard work anymore. Okay, maybe it was twice. Three, max!

Automatic sprinklers have put an end to a lot of science education. I learned, as a child, the effect of putting one's thumb over the end of the open hose to create a bottle neck. If the opening is made smaller, but the weight of all the water behind the bottleneck is pushing water through, the water must move faster - and farther. The evidence of bottle necks, under pressure, is well known and demonstrated with garden hoses.

But what if there is a bottle neck and very little pressure from behind? Well, of course that creates a dam and restricts movement. I learned about that from playing in an irrigation ditch. In the case of water, a lake is created. In my case, it also created a very angry grandfather. In the case of traffic, it creates a traffic jam. In the case of living organisms, it can restrict population size.

Animal and plant reproduction occur through time, like the flow of water. We often measure how long it takes for a natural process to occur and in what direction it goes. For example, we know how much water passes a given point each minute. But we don't always recognize the bottlenecks that affect the process between points of measurement. This is especially true in the biological sciences.

When we see a decline in the population of certain species of plants or animals, we usually respond by providing more food or restricting predation. Sometimes both, and either, of these can be helpful. But there are other requirements for a population of organisms to grow or remain healthy, and they often depend on bottlenecks in their timelines.

Take an infectious disease. To increase in a population, the organism that causes the disease must find a new, uninfected, non-immune organism to infect. Vaccination interrupts the number of available hosts for the disease, creating a bottle neck that restricts further infections. Vaccination has a public health role as well as an individual's protective role.

A study by Dr. Andrew Higginson at the University of Exeter in England found that, when nest sites are hard to come by, the species that suffer most are those that nest later in the year. He analyzed population changes in more than 200 bird species and 40 bumblebee species around the world and found that nesting sites and availability of food sources can both become bottlenecks. In other words, the early bird not only gets the worm, but it also gets the nesting site.

Native bees, such as bumble bees, the blue orchard bee, and many others often have specific nesting sites that are used from year to year. However, because of the places chosen and the native bee life cycle being somewhat invisible for much of the year, these sites are often obscure from our sight. They are then easily disturbed or destroyed by humans who don't even know they are there. The same is true of many small insects that are not well-studied or understood.

Long term scientific studies over extended areas are infrequently done. They are difficult, tedious, and expensive to conduct. In addition, many scientists have very short attention spans. My wife suggests that might just be me.

Anyway, we may find that the decline of insect populations has more to do with loss of nesting and breeding sites than with lack of food, predation, or even pesticides. As humans disrupt nesting sites without even knowing it, their actions become bottle necks for many animals. It may not be enough to sprinkle a little water on their heads and encourage them to grow.

The natives are restless

Long before the 1600's when the English established Jamestown, there were no honeybees in north America. There were flowers and fruits that needed pollinating, but no honeybees. Instead, most pollination was done by the lesser-known type of bee, now collectively called native bees. They are also sometimes referred to as solitary bees.

There are a great variety of these bees, but they are relatively little understood by the general human populace for two reasons. First fewer people than ever live or work the land where people might encounter them. But also, because of their unique life cycles they are far less visible on a day-to-day basis and are seldom aggressive so as to draw our attention. And the third reason they are a bit obscure is that they do not produce   directly.

Most people believe that the honeybee lifestyle is typical of all bee behavior. The truth is, the world is home to more than 20,000 species of bees, and a whopping 90% of them do not live together in hives. Instead, most of the world's bees are solitary, meaning they live alone. Each female solitary bee must gather pollen and nectar, build nests, and lay eggs all on her own, without the help of hundreds or thousands of workers.

Most insects have short life spans, even honeybees. Solitary bees also have a short active, adult lifespan of at best a few weeks. Male mason bees only fly for about two weeks—just long enough to mate—and females only live a few weeks longer. With such a short adult lifespan, solitary bees must use their time wisely! They do not have time to make honey or build elaborate comb nests.

They rise early, work late, and fly furiously to get things done. They do not like to fly too far from home, meaning they spend the bulk of their time preparing their nests and pollinating flowers within a relatively short period of time and over a compact distance. This makes them two or three times more effective as pollinators. However, they are far less far reaching in their overall effect. There efforts are often more important locally and are more difficult to exploit for economic effect.

> **Long before the 1600's when the English established Jamestown, there were no honeybees in north America.**

Life cycles

Without hives, where do solitary bees live? About 70% of solitary bee species nest underground in tunnels and burrows, while the remaining 30% nest aboveground, in holes in logs and stems. In either case, the female finds or digs a burrow, lays an egg, and then gathers pollen and nectar to deposit in a ball with the egg. She then typically seals the egg and food in the cell with mud or leaf debris, and then repeats the process until the borrow or tube is filled. If she has time, she may start a new burrow.

Interestingly, she can control the gender of each egg she lays and so lays female eggs first and male eggs later towards the opening of the tub. The males hatch a day or two earlier than the females and so open up the tube first. After allowing time for their cuticle to harden and perhaps finding a quick nectar meal, the males hang around the outside of the burrow until the female emerges and they mate almost immediately. When the females hatch a bit later, they find the tube has been cleared of their male siblings so that they can emerge also.

This seasonal timing means that many native bees are dependent on specific plants and flowers for their existence and are usually found in the vicinity of the same flowers each year. Because their nesting requirements and food supply are the same native bees are often found nesting in the same proximity each year, and at about the same time each year. This may give the appearance of their having a hive, but it is just a result of their life cycle habits.

The eggs hatch and the new larvae has a ready-made meal of pollen and nectar on which to feed. Eventually the larvae molts to form a quiescent pupa that does not feed and undergoes a metamorphosis to the adult stage. However, this whole process takes an entire year, so for the better part of the year that species of bee is not visible in the environment.

Exactly how the pupae know when it is the correct time to hatch is a bit of a mystery, but it is essential that the timing be coincident with the required flower blooms.

Life cycle significance

There are several ramifications to this life cycle that are interesting. One is that instead of being one solitary bee in an area there are usually a series of different species that show up throughout the year. These are usually somewhat consistent with the local flora blooms. Further it means that many solitary bees are seldom recognized or appreciated because of the short duration of their adult life stage. It also means that their nesting sites are often not visible or recognized.

This last issue is extremely problematic for the survival of many solitary bees. One of the greatest threats to their existence is human degradation of their nesting sites, usually without any knowledge that a nesting site was even present. Habitat destruction for construction of all kinds is harmful to an environmentally important insect and to human welfare.

The same habitat destruction also applies to the local flora and since each solitary bees relies of the blooms that correspond with the timing of the bee's emergence if the flowering plants have been destroyed the bees may also be doomed even if their nesting site is intact. It is unlikely that humans will cease construction and that does not bode well for the survival of many species of native bees, especially those not involved in the pollination of commercial crops.

Perhaps the only other solution would be to purposely increase the "wild" spaces available for the native bees to exploit. Many agriculturists already leave hedgerows, undisturbed areas, and wild plant spaces between fields, crops, and and/or roadways. However, the knowledge about solitary bees and an appreciation of their significance are not generally understood.

Further, even if such areas are widely established, it may take a long time to develop a new population as the bees only produce one generation a year, unlike honeybees that reproduce several generations in a season. Such spaces would need to be left in place, more or less undisturbed for long periods of time, a difficult task in the human controlled environment.

Management

Because of the complex environmental demands and unique life cycles it is much more difficult to manage solitary bees than honeybees or bumble bees to meet human needs. The hole-nesting bees have had the most success.

Two of the most common hole-nesting bee species used for crop pollination are the alfalfa leafcutter bees and the blue mason (orchard) bees. In the wild, both species nest in pre-made holes, such as old grub tunnels, crevices in peeling bark, or broken branches. As suggested by their names, leafcutter bees use pieces of leaves to build their nests, while mason bees use mud or clay. A small industry supplying both of these two types of bees has risen with some success.

Perhaps there are opportunities for other species attuned to different crops that bloom at other times of the year. That is not certain and would depend on commercial size operations based on those crops to make it viable.

However, the environmental role of the 20,000 species of bees cannot be ignored if we value the natural role these bees play in the environment. Other types of hole-nesting bees, such as sweat bees and carpenter bees, prefer to excavate their own holes in the ground, logs, reeds, or the dead canes of raspberry bushes.

Some reports indicate that nearly 40% of bees are facing extinction today, leaving many people wondering what they can do to help. Fortunately, the best thing you can do is to start local, in your own backyard. Making your garden as bee friendly as possible is as easy as adding things like native wildflowers and native bee nesting sites, including bee houses. Remember each seasonal species requires the seasonal flowers for food.

Like birdhouses, bee houses provide vital and otherwise missing nesting habitat. They are relatively simple in form, consisting of a birdhouse-like structure containing a series of exposed, reed-like tubes that the bees can lay their eggs in. Hole-nesting bees desperately search for appropriate nesting sites, sometimes even nesting in the ends of old garden hose nozzles, openings in metal garden furniture, or the hollow ends of wind chimes. Bee houses provide a more natural structure for the bees, and also allow for a bit of human assistance when necessary

On an annual basis, bee houses do need to be maintained and managed, or else they'll become uninhabitable. However, maintaining a bee house is simple: Just remove the bee-cocoon-filled nesting materials and store them in a cool place over winter. In the spring, remove the cocoons from the old materials and place them alongside new materials in your bee house. The new bee generation will emerge and get right to work pollinating your fruits and vegetable as well as your seasonal flowers.

THERE IS A SEASON

There is a season for cool damp evenings
With falling stars flashing in your eyes
And your voice saying anything

While we stand beneath the sky

There is a season for long hot summers
Among the brambles growing wild
Breathless while nothing stirred
Except my heart when you smiled

There is a season cool and colored
Shortened days darkened nights
Holding hands without a word
Reaping the harvest ripe

There is a season for winter morns
Watching smoke, snow, and ice
Staying with you where it's warm
Simple quiet seems paradise

There is a season for everything
When everything should seem to be
The season that is needed to bring
What comes next of everything.

Research ideas for Hymenoptera

We think we know a lot about bees. That is because humans have been keeping honeybees for centuries, perhaps longer. Yet, having been a beekeeper, I have discovered that a lot of what we think we know is more like an art than a science. Bees pretty much do what bees want to do. As explained, there are many types of bees and what we know of other types than honeybees are far more limited. I won't provide a numbered list as I have in the first couple of examples. But here are some general suggestions.

Honeybees

Because of the economic products and value in pollination services there has been a lot of research done on honeybees. However, a lot of it is directed at economic producers and problems. Honeybees suffer from numerous diseases and a lot has been done to study those. A lot of research has been about the individual bee, their diseases, or managing them for pollination and production.

However, a great deal more is needed. One of the drawbacks is that humans tend to look at the bee. The actual organism is the hive. Single bees cannot live outside the hive but for a few days at best. And there is still much unknown about the goings-on inside the hive.

Such research has been difficult because it is hard to observe inside the hive without disturbing the hive. It's like the uncertainty principle in physics. The act of measuring the position or speed of an electron alters the other thing you aren't measuring. In addition, any instruments inserted into a beehive, such as a thermometer or microphone is quickly discovered and plastered over with wax and propolis.

Modern technology is making it possible to monitor what is going on inside hives much easier and more accurately. Temperature measurements using infrared, sensitive microphones form outside the hive, weight changes, and even exhaust air analysis are now possible. How all these differ with season, time of day, bee subspecies, size of the hive, type of hive, and how any of it relates to hive health, life cycle, and other parameters are becoming possible. Another fascinating area of research open to creative naturalists is bee navigation, magnetic effects, aeromantic sensing, sound, and other natural parameters.

Native bees

The situation for native bees and other types such as bumble bees is far less developed. The numbers of native bees appear to be declining dramatically. This is thought to be due to loss of nesting sites, loss of flower and pollen sources, pesticide pollution, ecologic degradation, and general misunderstanding that they exist and are important. Further, most of them are not easily raised commercially so their numbers cannot be replenished as easily as honeybees.

For almost every region, simply finding out which ones are present at different seasons and times of the day would be a major contribution. Learning what their host flowers are, and their nesting habits and requirements would be major contributions also. There is a great need for public education and programs to protect their habitats.

One of the things that a dedicated amateur can provide than funded scientists generally have difficulty doing are long term studies. It is difficult to get funding for long term studies. But simply mapping and tracking the presence of species over the course of several years are invaluable studies on native bees. Especially where local observations of plants, weather, and environmental changes are noted because the local observer is a resident.

Such surveys can be done simply with sweep nets, sugar bowls, and examining individual flowers. Since native bees generally live as active adults for only a few weeks out of every year, it is important to make collections throughout the seasons and keep track over several; seasons of numbers and timing of their appearance. Know which plants they were collected on also gives clues as many flowering plants bloom only seasonally and the timing is correlated to weather and the emergence of their pollinators.

CHAPTER 9 - ODONATA

Phylum Arthropoda
Class Insecta:
Order Odonata:
Suborder Zygoptera - damsel flies

*"I am always surprised to discover
that when the world seems darkest,
there exists the greatest opportunity for light."*
Brigid Kemmerer

<u>The good, the bad, and the ugly</u>
   I have good news and bad news. Which do you want first? Wait! This is my book, so I guess I get to decide. The only reason I asked was that I'm not sure which news is good, or which is bad.
   The words "good" and "bad" do not have strict technological definitions. In fact, the two words probably describe a spectrum in which some things are perceived as good, but something else is perceived as either better or worse. They are values, not facts.
   There are things that most humans everywhere agree are good and bad. Most humans think killing other humans is bad unless it's for a cause that we think is worse. It seems most humans desire love and affection, although I sure can't figure out some people's taste. Why doesn't everyone love me?
   Science, of course, doesn't deal in values. Science deals with the physical world and matters that can be verified with empirical data. In fact, some scientists deny that there is any other kind of world and insist that consideration of things like religion, morality, love, or even beauty are bad. These are often the scientists who claim that we should fund scientific research because science does so much good.
   Obviously, the same science that brings us electric lights also created the electric chair. The science that

discovered antibiotics has created the antibiotic resistant bacteria that now plague us. The same science that created the nuclear bomb now provides us with news about nuclear catastrophes, using electricity generated by nuclear power plants.

It isn't the science that is bad. It's how the science is applied that can be considered good or bad. Some scientists like to attack religion as bad but fail to recognize that it is not the religious belief, but, just as with science, the application that may be faulty.

Did you know that generally a person with a common cold heals cuts faster than if they are not infected? So, are cold viruses bad or good? I suppose that depends on how seriously you are cut. If it is just a scratch, then no one cares, and the cold is miserable (another word for bad). If the carotid artery is severed, a cold is the least of your worry. But there might be an advantage to having a cold if you have a serious cut on the arm. The faster it heals the less chance of infection. Cold viruses can be good.

The parasitic ameba Entamoeba histolytica can make you and I deathly ill, but millions of people seem to harbor it with no apparent ill effects. So, is it bad or good? It just isn't clear.

I much prefer to be healthy. Healthy is defined as not having anything abnormal, or of order, or bad going on. Health is when everything is good. In fact, there are numerous things in my life that are bad: work, deadlines, meetings, broccoli. So, am I healthy? If I am not, then I must be unhealthy, or un-good. Isn't that the same as bad? It is if I am eating broccoli.

If I have a disease, I am not at ease, and that's not good. So, it must be bad. So, is broccoli a disease? Of course. But since bad is a relative term, there are some things worse than broccoli. No, I'm serious here.

Is high blood pressure worse than gout? People with high blood pressure are at risk of a heart attack and death.

Someone having a gout attack may not die; although they just might wish they could. Hydrochlorothiazide is a commonly prescribed water pill for lowering blood pressure, but it is one of the major causes of gout attacks. So, is Hydrochlorothiazide good or bad? Sorry! Science doesn't deal in value questions.

So, we have twenty percent of the American economy devoted to health care. Is that good? Well, I have good news and bad news. Which do you want first? The answer is ugly.

> **The words "good" and "bad" do not have strict technological definitions. They are instead opposites that exist on a spectrum in which some things are perceived as good, but something else is perceived as either better or worse. They are values, not facts.**

The beautiful and ugly

That brings me to the next categories: beautiful and ugly. Scientists are strangely obsessed with beauty. It's not that they are particularly beautiful, or that they any better judge of it than the average person. And I suppose it is a unique kind of beauty, maybe more like elegance of thinking than physical attractiveness.

Often you might hear a scientist say something about an "ugly experiment" or a "clunky conclusion". Scientists aren't big on expressing themselves well either. It's interesting to me that while many scientists dabble in art or music, seldom are they also poets. Mathematicians, who I'll include here with scientists as a professional courtesy, may speak of an "awkward" or "clumsy" proof. It's just all part of the scientist's inability to grasp concepts that may contain emotional content.

While this makes scientists socially clumsy, it is a benefit when it comes to doing science. You see non-scientists often confuse beauty with goodness and ugly with evil. While this error creates wonderful romance, adventure, and drama, it makes for poor physical reality and science. Scientists learn early that beautiful things can be quite dangerous and that some rather ugly things can be greatly beneficial.

How many women have been led astray by a "brown-eyed, handsome man?" I admit that beautiful women are less likely to be murderers than men. They tend more to psychological and emotional sadism. But to hear beautiful people tell it, they pay a high cost for their beauty in the form of expectations and social isolation. I wouldn't know.

For an example, proteins are essential to life but tend to be asymmetric lumps of amino acids with variable and amorphous shapes. However, they can function like little machines to pick up, set down, and move around materials within the cell. My son showed me a video of a small factory in China that made glass bottles. The machinery looked like it could have been used in a later 1800's industrial factory; dirty, clunky, awkward, and messy. That's how I imagine protein machinery.

On the other hand, the biological world is full of pretty, even beautiful, things, that are dangerous and vicious. For example, Damselflies, sometimes called Devils Darning Needles, are neither delicate damsel (at least not all of them them) or flies. Flies belong to the order Diptera, which mean two wings, because that is their unique characteristic. Damselflies have four wings. Nor are Damselflies dangerous predators of human souls nor do they do fancy needle work. They are beautifully and vividly colored, delicate insects that can hover gracefully in the air and dart about with the elegance of a dancer.

However, the order name, Odonata, means "toothed ones" and indicates that they are vicious and voracious predators, just like their cousins the Dragon Flies. Damselflies are beautiful, deadly carnivores. They have an extensible lower lip that can dart out to capture their prey. Even as babies they are carnivores that prey on other aquatic insect species.

However, you need not fear unless you are another insect such as a mosquito of other flying insect. They do not sting or bite larger animals. They do have some peculiar attributes, such as not being able to walk, despite having six legs.

Of course, maybe they can't walk because they have six legs. I know I can barely do it with two. Coordination can be difficult with that many appendages. Still, spiders walk with eight legs. Maybe one of you could figure all this out.

Beauty and the beast in human culture

I have no idea why humans are so fascinated by beauty juxtaposed around ugliness. From fairy tales, Phantoms in opera's, black widows, and praying mantises we seem to find ugliness and beauty, love, and violence, fascinating. While attitudes about damsel and dragon flies vary with different cultures, there are also many similarities in their perception.

The European impression of odonates has been that they are dangerous and malevolent; they are associated with snakes and the devil, as shown by many of their colloquial names. There are some seventy English folk names, 'Adderbolt' and 'Devil's Darning Needle' being two of them. Many western countries hold that they are snakes' companions. In America, a superstition was that dragonflies could stitch the mouths, and sometimes the eyes and ears, of lying children, scolding women, and cursing men. In European folklore it is sometimes suggested that Satan sent dragonflies into the world to cause mischief; in Italy and among the Dakota Indians, the insects are known as witches' animals. If the witch is the devil's creature, then, by association, so are dragonflies.

Asian cultures tend to see the Odonata differently. To the Chinese, they are an emblem of summer - but also a symbol of feebleness and instability. In Japan, they are revered and respected, being symbolic of happiness, strength, courage, and success. To the Japanese, the dragonfly is an important cultural symbol and was believed to be the spirit of the rice plant and a harbinger of rich harvests. Akitsushmi, which means Dragonfly Island, is an alternative name for Japan.

It isn't difficult to tell the difference between a dragonfly and a damselfly. Dragonflies have broad bodies and enormous eyes. Damselflies are slenderer and more delicate. When dragonflies are at rest, they hold their wings out like a glider. Damselflies will fold their wings over their backs when at rest. Dragonflies will often eat while in flight, and damselflies will always land before eating.

However, both are known for their fast flight and dazzling aerial feats. Because they are usually found near water, and they are active during the day their color and fluid motion mimics the movement of light on water. They can twist, turn, change direction in an instant, hover, move up or down, and even fly backwards. Dragonflies are sometimes known as mosquito hawks but, both the damsels and dragons are excellent hunter of flying insects. They can spot motion forty feet away.

Although some damsel flies have color pigments in their skin, for most the colors are caused by the refraction of light, like what happens with a rainbow. Structures and scales in their outer cuticle refract and reflect the light, often making them look iridescent green and blue.

When Beauty Is the Beast

When beauty is the beast
Death seems a delicate thing
Surrounded by iridescent colors
It arrives on transparent wings

Then the air hangs heavy
And the sun lifts light

Sparkling off the water
The blue beauty alights

Death folds her wings and waits
For the music only she can hear
To begin her deadly dance
That the careless fear

At length dancing with grace
Death hovers through the sky
Shimmering in the sun
Bloodthirsty damselfly

Research ideas for the Odonata

Who goes out looking for Odonata? Well, apparently, not very many people. And they haven't been looking effectively and persistently for a long time. It's a little like asking, "Is that the phone I don't hear ringing?" Because the Odonata are not economically important as crop pests, animal pests, or have commercial value there may not have been a lot of local information collected.

A local college, agriculture extension service, local fish and wildlife offices, or local parks and forestry offices may be of some help. Knowledgeable local entomologists can probably at least guide you in where to look for them. However, a local survey and collection of some kind is probably the first order of business.

But seriously, if you want to be a big fish in a small pond, the Odonata is a good place to start. Your friends will be fascinated. I am not saying in what manner, but they will find it fascinating that you are fascinated with the Odonata. If some of my suggestions sound repetitive, it's because I have a limited imagination. You can probably do better with a little effort and background research.

1. Numbers, kinds, seasons, and places need to be identified.

2. The same is needed for all life cycle stages.
3. Pigmentation and color pattern diversity in Odonata
4. Diets of different species
5. Water quality choices and effects
6. Wing maps
7. Effect of season, climate, and human population proximity on populations.
8. Parasites of dragonflies (internal or external)
9. Associated fauna
10. Effect of temperature on Odonata adults and larvae

As usual, long-term studies are almost non-existent. Funding for long term studies is difficult to come by and so the dedicated amateur can make important contributions to our understanding of nature over several seasons. Collections, field notes and written annual summaries and tables deposited in local libraries or extension offices could prove extremely valuable in the future.

CHAPTER 10 - ORTHOPTERA

Phylum Arthropoda
Class Insecta:
Order Orthoptera:
Numerous Families - Locust and Grasshoppers

"Even these of them ye may eat; the locust after his kind, and the bald locust after his kind, and the beetle after his kind, and the grasshopper after his kind."
Leviticus 11:22

Knowing verses telling

If you know something that would help others, but you don't tell anyone about it, are you responsible for the difficulties they may experience because they didn't know?

I've felt guilty for many years because I didn't offer "full disclosure" to my young, innocent daughters-in-law before they wed my sons. I suppose it was wrong. But by that time, it was either them or me that had to live with my boys; and I made my choice. I have since apologized. It isn't clear yet whether they have forgiven me.

This same question takes on a whole new meaning when we talk about the sciences. What if you know something that would cure a disease, but you refuse to tell anyone about it? Does that make you a murderer? What if you tell, but then charge an excessive amount for the cure?

What if you know some trivial little thing about the world that no one else knows, but you never tell anyone? That information can't be built on by someone else in the future because no one else knew or even knew that you knew.

What we mean when we say "biggest"

I read recently that the largest swarm of locust ever recorded was in the American west in 1875. This swarm reportedly covered 198,000 square miles and contained between 3 and 5 trillion locusts. They swarmed across the plains faster than a tornado and devoured all vegetation in their path. Witnesses say they covered the ground six inches deep and clogged rivers and lakes.

> **Old Testament**
> **Exodus 10:14**
> And the locusts went up over all the land of Egypt and rested in all the coasts of Egypt: very grievous were they; before them there were no such locusts as they, neither after them shall be such.

The question becomes, did any swarms of locust go unrecorded that may have been larger? In fact, how many swarms of locust of any size have ever been recorded? If grasshoppers swarm, but no one writes it down, did it really happen? When I was a kid, I once spent a week in Phoenix, Arizona in the middle of a swarm of locust. Did anyone record that one? I know I didn't.

I suppose I should write about my memories of living through a grasshopper swarm. It would have been best, scientifically, if I had written my experiences down at the time. However, I was only about fourteen and I find writing from memory easier than keeping a journal because you can lie.

Anyway, as I recall, the Phoenix plague was a lot of fun. We had to skim the pool every few minutes to swim in it because the grasshoppers continually covered the surface. Well, actually, I was the only one who would even swim in the pool at all. I suppose that says something.

Come to think of it, this need for writing down wisdom while we can still write, might be said about all kinds of things. If parents don't tell children what they have learned, the children can't build on their knowledge. Of course, whether children will bother to read what parents write is another question. Maybe if it's in a tweet.

Still, most people don't write much about their lives, wise or otherwise. They think no one would want to know anything about them or about what they have learned. Yet, most people would devour anything written about their ancestors if they found it.

I am sure good scientists like Galileo, Marie Curie, and Isaac Newton jotted down a few notes. Did they think what they had to say was important? Maybe only scientists should write stuff and everyone else should just be quiet. My wife says I have said enough.

It is sometimes even valuable to write things we don't know. For example, last week I wrote that I hadn't seen any insects in Sego Lilies this spring, and I wondered about pollinators. I got an e-mail from Bob Hooper in Montrose telling me that, on a recent hike, he "was reminded of that article." He ended up finding a variety of beetles and bees in Sego Lilies. Now we know, all because two people bothered to write something down.

> "I've learnt to gather simplicity from grasshoppers. I like their naive indecisive minds . . ., and I envy their ability to be able to mingle with the green..."
> **Munia Khan**

Quiet activity

At least writing things down is a quiet activity. There aren't a lot of that in the world anymore. Everything is audio or video, which is silly without the audio. Therefore, noise prevails. It is a lot noisier world than when people wrote things down.

The Orthopterans are some of the noisier insects. These group includes the crickets, but many grasshoppers can also make stridulations and other types of sound. Of course, it doesn't help an animal to make sounds unless others of the same kind of animal can hear it. Generally, making to much noise is dangerous for a little creature.

Anyway, then it should not be too much of surprise to learn that most Orthopterans have ears. You might find it a little strange to know that their ears are on their knees. Well sort of like their knee although with a many segmented legs it can be argued exactly which joint is the knee.

I think there must be some benefit to grasshoppers having their ears on their knees. They chomp their food so maybe having their ears a way away from their mandibles allows them to be less annoying to themselves. Doesn't make a lot of sense to us humans, I suppose. We just chew loud and ignore it.

Chewing

I'll have more to say about ears but right now I am distracted by the thought of Orthopterans chewing. Grasshoppers are herbivores, so a person's attitude toward grasshoppers depends a great deal on whether you grow plants or not.

Since children are generally not concerned with producing food, they may find them exciting and fun to play with. I found them especially useful as a young boy because most species will secret a brown foul liquid from their mouth when they are caught. We called it tobacco juice and girls were terrified of getting any of it on them.

If you raise crops grasshoppers can be devastating. They sometimes form huge swarms that can destroy a field of plants in hours. The largest grasshopper swarm ever recorded occurred was in 1875 in the Rocky Mountains. That swarm was 1,800 miles long and 110 miles wide. It's estimated that there were 3.5 trillion locusts in the swarm.

". . . blasted you with black blight and red;
laid waste your gardens and vineyards;
the locust devoured your fig-trees
and your olives;
yet you did not come back to me

Amos 4:9

However, they are not much different than other herbivores such as cattle. They are a necessary part of nature, consuming and in turn being consumed. John the Baptist is not the only person who has eaten locust. Many people in different parts of the world eat them as a rich source of protein. Like John, I would prefer mine with honey.

GRASSHOPPER ON GRASS

Now in the summer is there a leaf or blade of grass
Untouched?
Each has a blotched spot, chewed edge or such
These attest to the passage of time from bloom to tomb
As if the purpose of life is to produce for others to consume
What if the purpose of life is to consume so others can
produce?
Because there's no joy in living if producing is of no use

Now in the summer is there a consumer who will not yet be
consumed?
Each happy grazing insect will also meet its doom
Even the grasshopper on the grass chewing in the sun
Who leaps in startled abandon just to have some fun
What a lovely afternoon for a walk with lad or lass
Now in summer there's a grasshopper chewing on the grass

Research ideas for the Grasshoppers

Most folks think a grasshopper is a grasshopper. They assume size and color differences are age related. However, there are somewhere around 11,000 species in the world, and all share similar characteristics. The only way you can learn about your area is again through local experts and making your own observations.

More is known about grasshoppers and crickets than some insects because they can occur in large swarms that sometimes cause great damage to human crops. Even the few in your backyard can wreak havoc with your garden. But that doesn't mean that the amateur can't add significantly to our understanding.

Perhaps of less concern to humanity, but significant to the smooth operation of our world, grasshoppers are a major food source for many species of wildlife, and in some cultures even for humans. In addition, they sometimes serve as intermediate hosts for parasites of wild and domestic animals.

Other research areas might include:

1. Pigmentation and color pattern diversity in one species over a season
2. Spatial distribution of adults and larvae
3. Time dispersal of adults, larvae, and various species
4. Diets of different species
5. Water quality choices and effects
6. Wing maps
7. Effect of season, climate, and human population proximity on populations.
8. Parasites of dragonflies (internal or external)
9. Associated fauna
10. Effect of temperature on Odonata adults and larvae

Grasshoppers have had a long association with humans influencing art, food, as symbols, and in mechanical engineering. Some researchers have even trained grasshoppers to search for explosives.

"Of these you may eat every kind of great locust, every

> *kind of longheaded locust, every kind of green locust, and every kind of desert locust."*
> *Leviticus 11:22*

CRICKETS AND THE QUARTER MOON

Phylum Arthropoda
Class Insecta:
Order Orthoptera
Family Gryllidae - crickets

> *"When the cricket's song is the only sound you hear, how peaceful the whole earth seems."*
> *Marty Rubin*

Hearing noise

Apparently, the problem is I don't sit still. I think I am still. My mind is calm and I'm not using it much, so I'm told. But my wife says I fidget. I have to admit that I must because, every time she tells me I am, I find I am. I don't realize I'm doing it. I just do.

This came to my attention last week after I wrote about sound and made the statement that "When something moves it makes a sound." It follows that if I'm moving, I'm also noisy. I've admitted to popping knees when I stand up; but really, how much noise can a wiggling foot make?

The other problem is that she doesn't hear me when I answer her from the other room, so she thinks I didn't hear her. Does she think I'm deaf or something? I don't believe I realized how much sound affects my life until I wrote a column on it. I'm learning all kinds of new things, like how little irritations can lay dormant for fifty years.

Maybe it's the ears that are the problem. If you don't have ears, you can't hear sound. Not all animals have ears, you know. In fact, the number of animals that don't have ears far outnumber those that do. Of course, many of them are invertebrates, and you never hear much about them. It's probably just as well since we have too much noise already.

Lots of animals perceive pressure vibrations, but they don't necessarily hear sounds the way we do. Snakes, for example, are very sensitive to ground vibrations. They feel them through their lower jaws which, of course, are close to the ground since they must crawl on their bellies to move. That will teach them to harass Eve . . .

Interestingly, our hearing organs are situated close to our lower jaws also. Sometimes we can hear our jaws move. I can hear mine when I grind my teeth. But I only do that when I can't move anything else like my foot.

Crickets, grasshoppers, and other Orthopterans hear through ears that are located on their front legs just below their knees. Did you know that only male crickets Chirp? Female crickets move towards the male crickets' songs by "phono taxis". I tried using phono taxis when I was younger but either it doesn't work to well with human females, or I don't sing so good.

Interestingly, it's been only recently that scientists discovered that insects' ears work a lot like ours. They have ear drums that pick-up vibrations which are then amplified and passed on through fluid-filled tubes lined with tiny hair cells. These hair cells respond to specific wave frequencies in a manner much like our own hearing.

Crickets are also very good at determining the direction from which sound is coming from, better than humans. I can't ever tell where the stupid chirping is coming from at two in the morning.

It would seem awkward to have ears on your knees. If someone was saying something you didn't want to hear, instead of putting your fingers in your ears, you would just cross your legs and sing la-la-la-la.

Sound levels, in the environment, stay steady most of the time. Cricket chirping can be impressively loud in the fall; but as fall fades away, the Christmas carols start. So, the overall sound levels stay about even. There used to be a gap of a few peaceful days in late October, but that gap has since been filled with noisy, Halloween displays.

Potentially, anything that moves makes a sound. I wonder if cells make noise when they divide in two. Maybe they sound like the sound effect at the beginning of a TED talk. Anyway, I am trying hard to sit more still so the noise doesn't bother my wife as much. But the subsequent grinding of my teeth is giving me a headache.

> **The advantage of having ears on your knees is that if you didn't want to hear something you could just cross your legs.**

Making noise
    Another odd thing is that crickets sing with their wings. That gives a whole new meaning to soaring music. Of course, crickets really aren't particularly good at soaring. They mostly scamper or jump. I guess this is not too strange. We hear and speak with our heads so why shouldn't they speak and hear with appendages. Our heads are a sort of an appendage, sticking out on a skinny neck.
    Crickets chirp by rubbing their wings together. Male crickets have between 50 and 300 teeth on the left wing, like a comb. The right wing has a hard edge on the back. When the male rubs the left against the right it makes a sound. That sound is called a "stridulation", among other bad names depending on when and where it is heard. As usual, it's the males that create the commotion.
    Interestingly, while the male cricket song is necessary for his finding a mate, courtship can be dangerous. There is a parasitic tachinid fly, *Ormia ochracea,* that uses the song to locate the cricket host on which to deposit her eggs. Courtships always seem to be a little dangerous.

Consult the cricket
    "A loud cricket means that money is coming." No one ever told me that or I'd have been far more patient with crickets in the bedroom at night than I have been. I wonder how much money I have killed because of my impatience. No wonder I am poor.
    I am not sure I believe this though. I am told it is an old wisdom saying from China. I have nothing against the

Chinese, but from the pictures I have seen an awfully lot of them are poor. But the cricket culture in China dates back 2000 years and encompasses singing insects and fighting crickets.

Two thousand years ago China was amid the so-called Tang Dynasty. Contrary to my grandchildren's beliefs I am not that old, so I am just taking the word of others. But apparently at this time it was taken as remarkable that crickets could sing. I'm not sure why this was considered more remarkable than humans or birds, but it was. Anyway, it was then that people begin to capture crickets and keep them in little cages so they could hear them sing. There was no recording industry at the time so perhaps this is a little understandable.

In the song dynasty, from about 960 to 1279 AD the sport of cricket fighting became popular. It was much cheaper than lions and or gladiators, although the crowds were necessarily smaller also. Jia Shi-Dao became the Cricket Minister for nearly seventy years and eventually there was the Cricket Emperor, Ming Xuan-Zhong.

We know all this because of the major collections of poems, proverbs, songs, artwork, and newspaper clippings from the day. I guess the combination of singing voice, strength, and vitality were powerful attractants. I wonder why we don't ask our athletes to sing beautifully as well.

But many parts of the world revered crickets for different reasons. In Brazil crickets sometimes means impending rain or a financial windfall. A black cricket means illness, grey indicates money, and green indicates hope. I have never seen green cricket. Hmmm.

In other Brazilian cultures chirping crickets announces death if they chirp in the house. I think that one must be true because if they chirp in my house, it always announces their death. In other places it foretells pregnancy. A sure bet in poor countries.

Crickets have more recently been symbols of silence, whether as during a quiet evening or the embarrassed sort of quiet that might follow a bad joke.
Crickets are the most masterful musicians of the insect world. They have a variety of songs, each with its own

purpose. They have an invitation song to entice a crowd, preferably females. They have a courtship song and an after-wedding song where the male announces his bride. Males also sing songs to announce their territory and warn off invading males. Moreover, each individual cricket has its own unique volume and pitch.

There's no point to a cricket's song if it can't be heard, so both male and females have ears - on their front legs. Because their hearing organ is on their leg, they are especially sensitive to ground vibrations as well as sound. This is why they stop chirping as soon as you walk toward them, and then start again when you stop, or venture the other direction.

Insects as food

Today there is a great interest is raising crickets commercially for a variety of reasons. Many cultures routinely dine on insects and crickets are often a preferred type. However, more importantly insects are being raised as a food for yet other protein sources such as domestic animals, commercial livestock, and as fish food. Crickets are omnivores which means they can be raised on numerous different biological waste streams, whether plant of animal. Thus, they can be used both for decomposing wastes and as a further food source for other protein sources.

While much is known, and published, about how to raise insects much is still not known about the best, most efficient ways, foodstuffs, and other life parameters for all species.

Crickets and the Quarter Moon

I suppose of all the insects one could write poetry about
Crickets are the most likely, they are so melodious and all
They abound in literature, from the "Adventures of Pinocchio"
To "The Cricket on the Hearth" by Charles Dickens

Wordsworth and Keats have both immortalized crickets already
That alone is probably enough to make a poet forgo the attempt
Besides poetry about food is somehow distasteful
Oh yes, some people do eat crickets as unpoetic as that sounds
The sound of crickets has been associated with peace
I think the sound of a thousand crickets is more like a cacophony
Personally, I find their song monotonous and irritating
Especially in the bedroom in the middle of the night

Research ideas for crickets

Crickets are so common that many people think we must know all about them. That is the open door for the amateur scientist. Listed below are a list of questions that could beneficially be pursued by an amateur entomologist.
1. Seed preferences: since crickets will feed on plant material, they play a role in seed dispersal and consumption. Determining which seed they might consume, and their preferences in the wild would be important ecological information.
2. Crickets are well known to be parasitized by so called "Horsehair Worms", Nematomorpha, but the frequency, distribution, and course of such infections are not well understood. It is not clear how such infections might alter their normal behavior. In fact, the Nematomorpha are also not well understood, although they are, of course, not insects.
3. The effect of humidity and water intake, while understood to be essential, is not well understood on

the behavior and physiology of crickets. Changes in growth rate, limits, speed, cuticle formation and even hemolymph and blood cells distribution have not been studied adequately for many species.

4. There are questions about antennal response by crickets to approaching objects. For example, balls rolled toward them with minimal vibration seem to elicit antennal responses even though there is presumable little change in vibration and the antennae are not thought to be visual.
5. Maturation of the insect immune system has not been well-chronicled in many species of crickets. Especially the changes in cellular numbers and types.
6. The effect of cricket frass on soil ecosystems has not been studied. Nor has the effect of insect chitins effect on plants.
7. Since crickets can be raised in captivity relatively easily, comparing the result of insect chitin to crustacean chitin and its effects on plants and soil would be interesting.
8. Various species of crickets may have very different responses to temperature, humidity, light, and various forms of pollution. Seasonal distribution and spatial distribution of wild crickets are often needed.

CHAPTER 11 - LEPIDOPTERA

Phylum Arthropoda
Class Insecta:
Order Lepidoptera- moths and butterflies
Thin antennae with balls on the end - butterflies
Feathery antennae - moths and butterflies

"Well, I must endure the presence of a few caterpillars if I wish to become acquainted with the butterflies."
Antoine de Saint-Exupéry

Moths

Beautiful things in dark places have been an important theme in art and literature since the beginning of time. Come to think of it, ugly things in dark places have been equally fascinating. The ever-popular horror genre in art and literature are examples of these two concepts.

Normally science sheds light on the dark and dispels, or reveals, the horror. Scientists are often required to venture into the darkness of the unknown to discover truth, beauty, and occasionally the darker side of nature. It takes a special person to commit to a scientific career.

There is a week once a year when there is an opportunity for everyday citizens to experience the thrill of research and contribution to knowledge in one of the darkest topics of nature: the study of moths. The Global Citizen Science initiative sponsors the International Moth Week which usually occurs sometime in late July. It is with pride that I announce that this chapter was once part of an official event for International Moth Week. Information about other events can be found on their official website at http://nationalmothweek.org

It takes courage, intelligence, and a dedication to truth beyond that of normal human beings to venture into the dark to discover truth and beauty. Now, I don't want to brag. But when I told my wife that I had registered this column as an event during Moth Week, she-herself-told-me, "Gary, you're not normal."

People who study moths have been called by many names. Since moths are insects, such people are sometimes called entomologists. If these scientists specialize in only moths and butterflies, they might be called Lepidopterists after the name of the order Lepidoptera. Sometimes they are called "moth-ers", but that designation can lead to some unfortunate confrontations. Because they hang out in the dark, lepidopterists are also sometimes called witches or perverts.

A Moth Week is necessary because the great majority of moths are nocturnal, and they are not well known or studied. According to the folks at National Moth Week, there are between 150,000 and 500,000 species of moth. That's a pretty big range of numbers. It illustrates how little is known about moths. I know you've probably been outside in the dark, but were you looking for moths?

The roles moths play in our lives are somewhat obscure. This is because so much of what they do occurs after hours. Both adults, and their caterpillars, are food for a wide variety of wildlife, including other insects, spiders, frogs, toads, lizards, shrews, hedgehogs, bats, and birds.

Moths have a great impact on plants because they eat their leaves and fruit. This damage is limited in that many types of plants have evolved to produce special chemicals that fend off these attacks and limit the damage. Interestingly, the chemicals evolved by plants have not been aggressively explored for their use to protect crops.

At the same time, moths also benefit many plants by pollinating flowers while feeding on their nectar, and so help in seed production. They benefit not only crops, but many wild plants as well. This is sometimes an overlooked issue in environmental concerns. There are a lot of overlooked issues at night.

Moths play a vital role in telling us about the health of our environment, like the canary in the coalmine. Since they are widespread, found in many habitats, and are sensitive to change, moths are particularly useful as indicator species. Monitoring their numbers and ranges can give us vital clues how changes in our activities in farming, mining, or recreation can have an overall environmental affect.

You can learn more about how you can help monitor moths at The National Moth Week web site. They also have a page on how to find moths and report your findings. Lastly, let me assure you that moths will not wet in your hair and make your hair fall out like my wife's grandmother told her. My wife's grandmother's entomophobia has caused all kinds of problems in my career.

> **Beautiful things in dark places have been an important theme in art and literature since the beginning of time. Come to think of it, ugly things in dark places have been equally fascinating.**

The number of moth species outnumber butterflies nine to one. They are generally less appreciated because most are nocturnal. Color perception is difficult at night so many of them are rather drab in color, compared to butterflies. However, moths are significant pollinators of some plants, especially desert varieties.

Many moths do not feed as adults but live on stored energy acquired as larvae. Moths presumably navigate by either starlight or moonlight although it isn't clear and much remains to be understood about this. But such a reliance on stars or moon may be the cause of their being attracted to lights and flames.

Moths have a highly developed sense of smell. The odor receptors are located on their feathery antennae and is presumably the reason they have such filamentous antennae. It increases the surface area for sensing chemical signals.

FLAMING MOTH
He's a plain brown wrapper,
But there isn't any shame.
No need for color in the moonlight.
He tumbles in just before the rain,
Slipping and sliding he came,
Keeping out of sight.

EVERYTHING THAT CREEPETH ON THE EARTH

In some ways he's like a man of the cloth,
No one even noticed he came.
I know you think you should bet on the moth
But I think I'll bet on the flame.

Searching for an evening flower,
Her scent is in the air.
Hiding on the bark of a tree and
Moving gently with care.
Fluttering from here to there,
Twisting in the evening breeze.

In some ways he's like a man of the cloth,
He acts as if this were a game.
I know you think you should bet on the moth
But I think I'll bet on the flame.

Keeping one eye on the silver orb,
The light directs his flight.
Sending feelers out into the dark,
Alert for a night-long fight.
Hampered by many faceted sight,
His path an earth curved arc.

In some ways he's like a man of the cloth,
He can't help it, he's not to blame.
I know you think you should bet on the moth
But I think I'll bet on the flame.

He follows the distant and ancient glow
It marks eternal flight.
It has served him for a million years.
Yet the closest star is the most bright,
A blazing flame in the dark night.
Though that way ends in tears.

In some ways he's like a man of the cloth,
Blinded by the light of the flame.

*I know you think you should bet on the moth
But I think I'll bet on the flame.*

Research ideas for moths
    Moths in general are not as well known or understood as butterflies because they are active after dark and so less commonly observed and studied. However, they are still significant pollinators and important areas of study.
    Many of the same things can be investigated as for butterflies. However, there is the added difficulty that most are nocturnal. However, many of their immatures are active in the day and this can make the study of immatures easier. Presumable, when the young and the adult are separated by time, space, and food type it reduces competition between the generations for resources.
    Simply sampling species by time of night when they are active is a significant task. Comparing locations in this manner is also useful.

Butterflies

Phylum Arthropoda
Class Insecta:
Order Lepidoptera:
Family Nymphalidae:
Danaus plexippus - Monarch butterfly

"Butterflies can't see their wings. They can't see how truly beautiful they are, but everyone else can. People are like that as well."
- Naya Rivera -

Monarchs
    But while you are studying the Lepidoptera don't eat the Monarchs. They taste terrible and would probably be poisonous if you chocked down too many. This is because they feed exclusively on Milkweed as immatures and their adult body Is filled with cardenolide aglycones which is a heart

arrestor in large enough doses. In smaller doses it might just make you vomit.

Besides, they are dwindling in number very quickly. Since about 1996 their numbers have declined from around one billion to about eight-hundred million. They have enough trouble because of what they eat without being further damaged by being eaten.

There is something beautiful about close, intimate relationships that support both members of the relationship. It's like long successful marriages. Of course, like long marriages, once one member of the relationship disappears, the other usually isn't long for this earth either. However, don't get too romantic about close intimate relationships. Sometimes things are a little different than they appear.

The plant

For example, it might be tempting to imagine the milkweed produces cardenolide aglycones just for the good of Monarch butterflies. Actually, the plant produces these to ward of predators exactly like butterfly caterpillars. Poison is just one of its arsenals of defensive mechanisms. Birds and grazing animals that take a nibble often find themselves very sick and learn quickly to leave the plant alone.

Milkweed also has hairy leaves which also deters a lot of herbivores. I don't think I have ever had a salad with hairy leaves unless it was from the hair of the cook or waiter. It doesn't sound appetizing. Most grazing animals agree.

Then there is the latex. When milkweed leaves are damaged, they exude a sticky latex like material that is loaded with the cardenolide aglycones. It's pretty hard to consume the plant with getting all sticky and gooped up. Smaller insects that may attack the plant be become enmeshed in the sticky goo and be unable to escape. Some tiny stages of caterpillars can even have their mouths stuck shut so the starve to death.

So, you see, this seemingly benign relationship is actually about a plants ability to ward off consumers and a specific consumers behavior that allows it to attack the plant anyway. Not as romantic sounding I suppose. Sort of like a long, successful marriage.

The Butterfly

You might wonder, how does the Monarch deal with all these problems? Contrary to common belief it appears that monarch caterpillars can and do suffer the effects of consuming cardiac glycosides. Different species of milkweeds, or even different individual plants within a species, can vary significantly in their cardenolide levels. Caterpillars feeding on milkweeds with highest levels of cardenolides have lower survival rates.

Therefore, female butterflies prefer to oviposit their eggs on milkweed plants with lower cardenolide levels. If the ingestion of cardiac glycosides were completely beneficial to their offspring females would seek host plants with the highest toxicity.

Monarchs' tolerance for "hairy" salads is a little difference. It turns out they don't like them anymore than you or I. However, the early larval stages (caterpillars) are very adept at shaving the hairs off in a little circle so they can then munch down on the plant. Some caterpillars have been observed chewing a ring in the middle of the leaf so that the latex oozes from the damaged area while they gleefully chew on the center of the ring. It's sort of like trenching your tent for a rainstorm.

Older, larger caterpillars will sometimes chew a notch in the leaf stem, and at the tip of the leaf. This causes the leaf to droop, and the latex then drains out at the tip.

The Pair

What appears to be a mutually beneficial arrangement is a war between defensive and offensive behaviors of the two species. The problem is the monarch's life cycle depends on this specific plant for their survival. But milkweed habitat in their summer range has been increasingly disrupted by human activities such as agriculture and urbanization. There is also a similar problem in the migrator's destination as well. The acreage of their breeding sites has been reduced from a total of about fifty acres to less than two acres.

In recent years some species of the milkweed plant has developed the ability to grow new leaves quickly. If the plant becomes too severely denuded by the Monarchs, new tine leaves begin to show up quickly.

The poem refers to a play on words for *Asclepias*, which is the genus name for milkweed and *Asclepius* who was the Greek God of medicine. The symbol for Asclepius was two serpents entwined around a staff. In turn, the two serpents are thought to be symbols of the earliest known treatment for any disease. This refers to what is presumably the ancient method of removing *Dracunculus medinensis* from skin lesions referenced in the Old Testament as serpents on a cross. The story is related in the Old Testament about an episode about the Children of Israel wondering in the desert.

MONARCHS AND MILKWEEDS

It seems a sleepy transformation
But things are seldom as they seem
Like the snake entwined rod
Of mythology's theme
One life entwined with another
Such that neither can escape
So that throughout all the changes
Each determines the others fate

Consider Asclepias and Danaus
Like many ancient Greeks
Where each one thrives on each
With offense they turn the other cheek
And like the stories of myths and legends
There are lessons for us to learn
We must travel far for the things for which we yearn

Research ideas for Lepidoptera

There are two main groups of the Lepidoptera: the butterflies and moths. Butterflies are generally active during the day and moths are nocturnal. There are other differences in their sense organs and coloration due to their differing habitats. I have discussed Monarch butterflies as they are well known, but many lesser-known butterflies await information to be gathered about them.

Because butterflies are highly visible and ubiquitous humans have studied them for a long time. There are around 18,000 species of butterflies in the world and around 750 are found in north America. In fact, amateur collectors of Lepidoptera have collected more butterfly specimens than all professional scientists have since the 1800's. This has been a tremendous boon to entomology because over a million of these collections have made their way into public databases.

However, the number of collections and collectors has declined sharply since the 1990's. This declining acquisition has seriously compromised important testing of current and future ecological hypotheses and hindered efforts to forestall the decline of insect populations worldwide. Obviously the first order of business might be to encourage responsible, local lepidoptera collections and see that they are placed in public repositories.

Raising Lepidoptera is also a challenging hobby that requires a lot of knowledge and trial and error. Many wild species have never been raised in captivity for a variety of reasons. Being able to do so consistently would be extremely valuable knowledge for the world.

LUNA

Phylum Arthropoda
Class Insecta:
Order Lepidoptera:
Family Saturniidae
Species *Actias luna* - Luna Moth

The Luna moth has two spots on its lime green wings, often encircled by colored bands. Therefore, they are placed in the Family Saturnidae, after the rings around Saturn. Linnaeus named the species after its spots also, calling it luna which means moon.

Luna moths have "tails" on their wings that are crinkled and don't lay flat. Apparently, bat echolocation is slightly confused by these rumpled tails, and they snap at the tail instead of the moth. This gives the moth a fraction of a second to escape. Luna moths can fly without their "tails," but bats are much more likely to catch them.

Because they are moths and nighttime flyers they are seldom seen. They are usually found in wooded areas and may not show up without a full moon and clear skies.

LUNA

Have you ever seen a moonbeam?
I have and it was green
It was like a wave oscillating in the air
Like a lover's unspoken prayer

Have you ever seen a star shine?
With the moon beams all entwined.
Like a slowly glowing isotope
Gliding up and down a gentle slope

Have you ever seen a moth at night?
Glowing green under moonlight?
Then you know what music means
When it soars all shiny green

"Provide for yourselves purses that do not wear out,
and never-failing treasure in heaven,
where no thief can get near it, no moth destroy it."
Luke 12:33

CHAPTER 12 - DIPTERA

Phylum Arthropoda:
Class Insecta:
Order Diptera:
Species Musca domestica

*"The masses of flies over the dirt do not state their unity;
it is the dirt that brings them together."*
M.F. Moonzajer

<u>Death and flies</u>

Flies! They have been on my mind lately. I suppose that's better than my mind being on some of the thing's flies are normally found on. Or maybe that says something about my mind. I'm not sure which, so I'll let my wife straighten that out when she edits this book.

We were driving a granddaughter on a road trip once and discussing her education plans. I was advocating for more math and science over literature because it had so many practical applications. As part of the effort to sway her, my wife read aloud "The Microbe Hunters" by Paul DeKruif.

Great book, by the way. It was voted "the most influential book to becoming a scientist" in a poll taken a few years ago by the scientific research society Sigma Xi members. I think that was supposed to be an honor. Maybe it was a warning.

Anyway, back to the road trip. On one of our stops, a fly got into the car. My wife, being the sweetest thing in the car, was being pestered to death. Suddenly, with an accurate and well-timed blow, my wife used the book to terminate the fly. I remarked how that was evidence of the practicality of science, but my granddaughter countered that she thought the book was more like literature.

I am not sure exactly what kind of fly was in the car because the remains were too damaged to make an accurate identification. But there are a lot of different kinds of flies. Let's see, just to name a few of the fifty thousand, there are houseflies, blow flies, horseflies, fruit flies, stable flies, black flies, tsetse flies, dear flies, gadflies, dung flies, robber flies, sand flies, louse flies, bee flies and mosquitoes.

Oh yeah, mosquitoes are part of the flies as well. Insects can be roughly classified into those with two pair of wings (the most common condition) those without wings (far fewer in number) and those with only one pair of wings, the flies, or Diptera (two wings). Mosquitoes have only one pair of wings, so they are filed as flies because of how they fly.

I learned about mosquitoes as flies back when I started doing research on them. I was pretty lucky because a lot of young biologists are overwhelmed and don't know how to choose a field of study or an experimental animal. There are several theories about how to decide on a research subject.

Some scientist's council students to follow their passion. If more scientists would do this, there would be fewer scientists because it's pretty hard to get passionate some aspects of science. Besides, this approach has already given us too many wolf, bear, lion, and shark experts. The other popular, passionate categories are the cute little animals like foxes and chipmunks. It's hard to work up passion for worms, flies, and mosquitoes. Even those of us who might possess such passion aren't going to admit it.

Another approach is the one I used when I let the species choose me. As I recall, it was a hot, muggy day on the lake. A bunch of us from the lab were using the University research watercraft (oh alright, it was a large rowboat) to collect snail specimens. (Don't ask.) As we barreled into some cattails, a cloud of mosquitoes erupted around us, and it soon became evident that I was their favorite target.

There you have it. Why spend money on research animals when they can be collected for free by me just standing around? Actually, I can collect even more efficiently if I lay around. I tell the wife I have to go to the office, then head for the hammock.

Of course, there is another method of choosing research subjects. Some scientists advocate choosing subjects that have the potential to making money from your research. I sure wish I had had that on my mind instead of flies.

Anyway, I once caught a fly by its wing. That might be useful rare information for a book like this. Actually, I caught the whole fly in my hand, but I wasn't sure it was there. So, I opened my hand just a little, and the dazed fly crawled out onto my thumb. I pinched its wing between the thumb and finger of my other hand, and there you have it, I had caught a fly by its wing.

When this happens, the fly becomes alarmed and tries very hard to fly away. This attempt to fly away creates a very loud buzz. I don't know if the buzz is any louder than usual, but it seems especially loud if it occurs during a very quiet moment of a church service.

How was I to know? I was only ten. My rib still hurts at the memory of the elbow my mother gave me. And the blow made me lose the fly. I wonder what the statute of limitations is on child abuse...

At least I wasn't pulling legs or wings off the fly like some kids. I think that difference is what kept me from becoming a serial killer and becoming a biologist instead. It's amazing that from that early beginning, I have spent so much time with flies. Maybe it's not so strange when I think really about it. There are over 50,000 kinds of flies in the world, so it's pretty hard not to spend time with them.

They are amazing little creatures if you can examine them without much thought as to where they may have been recently. If you can't do this, you are normal. If you can do this, you are a biologist. And you thought you needed a degree for that. No, you only need the degree to get paid as a biologist. You can be a biologist for free all you would like.

> *"Dead flies make the perfumer's sweet ointment*
> *turn rancid and ferment;*
> *so can a little folly make wisdom lose its worth."*
> *Ecclesiastes 10:1*

Flies can do a number of amazing things. Like fly, for example. Bet you can't do that, at least not the way flies do. Their wings serve two functions, as both wing and propeller, simultaneously. To accomplish this, they must flap their wings about 200 times a second. Rattlesnakes only rattle about 100 times per second, and hummingbirds flap their wings only about 75 time per second. The best a human can flap anything is about ten times per second.

Understanding how flies fly can be very useful knowledge in some settings. Because a fly has to gain altitude before it can move forward, it always takes off straight up, perpendicular to the surface it is on, like a helicopter. It first rises vertically for a couple of inches before moving forward. Knowing this, if you clap your hands from the side about three inches above the fly, you will almost always catch it between your palms. With a little practice, you can catch a fly one handed using the same strategy. Then simply throw your catch onto a hard surface to stun it and prove that you caught it barehanded in mid-flight.

Imagine how impressive that kind of dexterity would appear to all the girls around you on a picnic. In reality the feat is accomplished by your superior knowledge of how living things function, but there is no need to explain that. Just let everyone assume you have superior physical abilities. That's what seems to be important to the girls anyway. Girls can be so shallow.

Another thing a fly can do is land on the ceiling. I bet the Air Force would give a lot to know how to do that! Does it do this by executing a half roll or an inside loop? Ha! Much more interesting. The fly simply flies close to the ceiling in a normal fashion. Then it reaches up with its two front legs, grabs the ceiling with its little feet, and somersaults into position.

You learn these things by careful observation during church services.

Class II medical devices

Maggots, leeches, scorpions, toads, newts, bats, lizards, goats, and spider venom: "by the pricking of my thumb something wicked this way comes." Of course, the "wicked something" in the quote was a human; Macbeth from Shakespeare's tragedy of the same name.

From my study of animals, I have concluded that the most frightening are humans. All the other kinds are just part of nature, and they often provide us with surprising benefits. Humans, on the other hand, do some really despicable things.

Most people would say that maggots, for example, are disgusting, although not necessarily wicked. Blowflies are a group of large flies that are brightly colored in metallic blues, greens, and coppers and are actually very pretty when examined closely. Blowflies tend to lay their eggs in dead animal tissue. The immature (the Anti-defamation League suggests the change in nomenclature) consume the flesh and, thereby, aid in the decomposition of dead animals.

It is hard to imagine what the world might be like without the help of blow fly maggots, er, immature. (Knee deep in rotting carcasses I presume.) Blow flies can even be handy in determining how long a living animal has been dead. The flies are cold blooded animals and develop at a given rate at known temperatures. They also pass through a set sequence of recognizable stages. Therefore, their level of maturity in the corpse can indicate how long the body has been dead.

Oh, but there is one tiny problem. Some of the flies lay their eggs on healthy, living tissue, and the immature literally eat holes in the skin and consume living flesh. This condition can be extremely costly in livestock as it sickens animals and ruins hides. This condition can also affect humans where it is called myiasis. (By the way, it is unseemly for people who have watched "Nightmare on Elm Street" or "Night of the Living Dead" to get all fastidious on me here for describing the natural life cycle of a blow fly.)

Maggots, er, immature of the common Green Bottle Fly eat only dead tissue and I don't know why it is called a Green Bottle Fly. Is it a green fly, or was it found in a green bottle? Maybe it's a green fly that was found in a bottle. This is a serious advantage in some cases. No, I don't mean the fact that it is found on bottles, because it isn't. The advantage is it eats only dead tissue.

It was noted centuries ago that some maggot-infested wounds (Oh, forget it!) of warriors on the battlefield healed more quickly and cleaner than other wounds. In the American

Civil War, a human, Dr. John Zacharias used maggots to remove dead tissue in gangrene cases, as he said, "with eminent satisfaction." Whose satisfaction, I wonder. That's just gross.

The practice of maggot debridement has suffered over the years for aesthetic reasons, but recently it has been resurrected. Perhaps revived would be a better term. Anyway, it seems that with the development of antibiotic-resistant bacteria, wound debridement by surgical maggots has again shown some advantages. Not only does maggot debridement yield a cleaner wound, but there also appear to be other positive factors associated with maggot infested wounds. Maggots may have growth-stimulating effects on patient tissue as well as antibacterial effects.

Dr. Ronald Sherman, University of California, Irvine, (yet another human) has received FDA approval to produce and market surgical maggots. Yes, he grows tissue-infesting maggots in a lab to sell to other humans, called doctors, to purposely embed in wounded patients! The lab-produced, disinfected maggots are called "non-exempt, Class II medical devices with special controls." Immature? Class II medical device? A maggot by any other name is still a maggot. This is what happens when political correctness runs amok.

I'm telling you; humans can be positively terrifying. "By the pricking of my thumb, something wicked this way comes."

Well, Spiderman might be able to climb walls, but flies can walk on ceilings. Yet I don't think anyone wants to be Fly Man. House fly taste buds are on their toes. That's why when they land on your birthday cake, or anything else, they immediately start walking around to see if it tastes good. The problem is they aren't very discriminating. While they like your cake just fine, they also quite enjoy feces and rotting food as well, and that may have been where they were just moments before your cake.

Then there is the problem that they have no teeth. They lie on a liquid diet. So, another problem develops once they stop walking. When they stop, they are probably regurgitating digestive enzymes onto your cake so they can slurp the juice back up.

A liquid diet does have its own set of problems. Food passes through the digestive tract very quickly. This means that nearly every time a fly's land on food it also poops, if not immediately, almost instantly when it finishes eating.

Overall, I think flies are not desirable houseguests.

> **Old Testament**
> **Exodus 8:21-22**
> I will send swarms of flies upon thee, and upon thy servants, and upon thy people, and into thy houses: and the houses of the Egyptians shall be full of swarms of flies, and also the ground whereon they are. And I will sever in that day the land of Goshen, in which my people dwell, that no swarms of flies shall be there; to the end thou mayest know that I am the Lord in the midst of the earth.

LIES ON FLIES

You'll probably think it's pack of lies
When I tell you how many flies
That this world abides
It's enough to make one cry
There are 120,000 kinds of flies
But just a pair of mating flies
With nothing around to cause demise
In only six days of mating tries
Can produce a quintillion flies
Equal to the mass of the earth in size

Mosquitoes are flies
Are you surprised?
They have just two wings which is why
Sciomyzidae and Sepsidae are also flies
Just two wings and compound eyes
Is always enough to qualify
That, and six legs, make a thing a fly

There are Horse, Deer, and Crane flies
All named after animal lives
Black, Yellow, and Blue Bottle flies
Are named for colors they apprise
I'll let you be the one to decide

Why there are Face, Flesh, and Horn flies
Sand, Stable and Shore flies
One can predict where they apply
But it is impossible when you realize
You can't name all the kinds of flies
So I don't think I'll even try

CULICIDAE

Phylum Arthropoda
Class Insects:
Order Diptera:
Family Culicidae - mosquitoes

    The next time you have electrodes planted in your brain, thank a mosquito. Electrodes must be very tiny to slip through the skin with little irritation. Kind of like a mosquito proboscis. That's what the female mosquito mouth part is called. But you don't have to wait. The next time you have a shot or give blood you can thank a mosquito. Same thing. Scientists have studied the mosquito proboscis to determine how to make invasive procedures such as these more effective, less dangerous, and less painful.
    Most people are not properly grateful for these contributions and the mosquito may be one of the most hated insects. However, I can tell you, if you have a problem with mosquitoes, it is of your own making, or one of your neighbors. Mosquitoes are simply poor fliers. Top speed is only 1 to 1.5 miles per hour. Further, most mosquitoes never fly more than a couple of miles from their breeding site. The Asian Tiger Mosquito can only fly about a hundred yards. So if they are in your yard, but not breeding there because you have no

standing water, the problem is probably within a mile or so from your home.

That gentle, deadly buzz you hear is caused by the mosquito wings, which beat between 300 and 600 times per second. Interestingly, the female apparently hears and responds to the male's wing beats, and as they approach for mating, she synchronizes her wing beat with his. Males don't take blood meals and live peacefully imbibing plant nectar. So the dreaded hummmnnn you hear in your ear is always a female. They say the female is always the most dangerous of the species.

Anyway, I got to thinking about mosquitoes. I suppose it was because of this book, but I can never really tell why I get to thinking about things. But I think about mosquitoes a lot. There was no particular reason, other than a fear for my life. I don't think exsanguination is particularly painful but itching to death is an awful way to go. It's messy too. See, I killed mosquitoes professionally for almost twenty years. There is a contract out on me.

Humans have made a lot of advancements in mosquito control in recent years. You can't tell because we have also regulated most of it into inefficiency. But recent breakthroughs are promising - of something. I'm not sure what.

There are a number of products that kill mosquito larvae at extremely low doses and cause no discernible environmental problems. However, these products have to be registered for each specific use, and the testing that is required by the government is costly. The mosquito larvicide industry is small so the companies could not justify the testing. The products are still available and used extensively in other areas of application. They just can't be used for mosquitoes anymore. Thank goodness for the benevolent eye of the bureaucrats.

I became concerned, again, when I recently learned that I must have permission to kill mosquitoes if I am not on my own property. I was about to slap a mosquito on my friend's arm as we visited the other day, and he warned me that he was told it was against the law to kill mosquitoes on other people's property - unless you had written permission. I thought he was just making that up because I have, on occasion, killed a mosquito on him a little more violently than he thought necessary. He assured me it was so because the local mosquito control district said so.

What? People kill my bees with their pesticides all the time without my permission simply because my bees are on their property. So, I can't kill the mosquitoes raised on their property when they come onto mine without permission? Someone needs to organize and get united like the Republicans and Democrats. I just can't figure out if that should fall to the bees, the beekeepers, the mosquitoes, or the mosquito breeders. I'm also not exactly sure who to unite against. This conundrum is an excellent metaphor for life.

The problem with trying to regulate biological systems, whether it's a disease vector like mosquitoes or disease recipients like humans, is that living things operate according to internal and environmental cues much more readily than they do regulations. For example, it is pretty difficult to arrange animal reproduction so that birth is guaranteed to occur when we regulate it. Mosquitoes tend to hatch when conditions are correct, not when the regulations have all been met.

I spent a lot of time, once, mapping something called degree days to try to predict mosquito populations. The best I could ever do was to predict hatching within about a two-week period. Then it dawned on me that I could predict that accurately just by looking at the calendar. Looking at the calendar was a lot easier than calculating degree days, so we stopped doing it. Both ideas are based on numbers, I guess, so they are equally scientific.

The problem is that government bureaucrats want to follow all the rules, even when, by the time the rules are followed, it is too late to apply them. Regulations seem to work better on inanimate objects like buildings, vehicles, pipes, and wires. Humans and mosquitoes don't regulate well.

Recent research suggests that we may be able to eradicate mosquitoes using genetic manipulation. There is hardly a doubt among geneticists that it could work. Geneticists seldom have any doubts.

If they are correct there are just a few remaining questions. Would it be ethical? Let's see, we expend great effort and expense at preserving endangered species. We lament the several hundred species we know mankind has already eliminated. So let's wipe another one out on purpose?

The last question must be, "Would there be any unforeseen consequences?" Gosh! Could there be?

> **The next time you have electrodes planted in your brain, thank a mosquito.**

Mosquitoes

So, anyway, I got to thinking about mosquitoes. There was no particular reason, other than a fear for my life. I don't think exsanguination is particularly painful but itching to death is an awful way to go. It's messy too.

Humans have made a lot of advancements in mosquito control in recent years. You can't tell because we have also regulated most of it into inefficiency. But recent breakthroughs are promising - of something. I'm not sure what.

There are a number of products that kill mosquito larvae at extremely low doses and cause no discernible environmental problems. However, these products have to be registered for each specific use, and the testing that is required by the government is costly. The mosquito larvicide industry is small so the companies could not justify the testing. The products are still available and used extensively in other areas of application. They just can't be used for mosquitoes anymore. Thank goodness for the benevolent eye of the bureaucrats.

I became concerned, again, when I recently learned that I must have permission to kill mosquitoes if I am not on my own property. I was about to slap a mosquito on my friend's arm as we visited the other day, and he warned me that he was told it was against the law to kill mosquitoes on other people's property - unless you had written permission. I thought he was just making that up because I have, on occasion, killed a mosquito on him a little more violently than he thought necessary. He assured me it was so because the local mosquito control district said so.

What? People kill my bees with their pesticides all the time without my permission simply because my bees are on their property. So I can't kill the mosquitoes raised on their property when they come onto mine without permission? Someone needs to organize and get united like the Republicans and Democrats. I just can't figure out if that should fall to the bees, the beekeepers, the mosquitoes, or the mosquito breeders. I'm also not exactly sure who to unite against. This conundrum is an excellent metaphor for life.

The problem with trying to regulate biological systems, whether it's a disease vector like mosquitoes or disease recipients like humans, is that living things operate according to internal and environmental cues much more readily than they do regulations. For example, it is pretty difficult to arrange animal reproduction so that birth is guaranteed to occur when we regulate it. Mosquitoes tend to hatch when conditions are correct, not when the regulations have all been met.

I spent a lot of time, once, mapping something called degree days to try to predict mosquito populations. The best I could ever do was to predict hatching within about a two-week period. Then it dawned on me that I could predict that accurately just by looking at the calendar. Looking at the calendar was a lot easier than calculating degree days, so we stopped doing it. Both ideas are based on numbers, I guess, so they are equally scientific.

The problem is that government bureaucrats want to follow all the rules, even when, by the time the rules are followed, it is too late to apply them. Regulations seem to work better on inanimate objects like buildings, vehicles, pipes, and wires. Humans and mosquitoes don't regulate well.

Recent research suggests that we may be able to eradicate mosquitoes using genetic manipulation. There is hardly a doubt among geneticists that it could work. Geneticists seldom have any doubts.

If they are correct there are just a few remaining questions. Would it be ethical? Let's see, we expend great effort and expense at preserving endangered species. We lament the several hundred species we know mankind has already eliminated. So let's wipe another one out on purpose?

The last question must be, "Would there be any unforeseen consequences?" Gosh! Could there be?

> **I don't think exsanguination is particularly painful but itching to death is an awful way to go. It's messy too.**

When I worked in mosquito control, we mostly attacked the larval stages. It's easier because they are restricted to specific areas and once you find the area it's much easier, more economical, and efficient to kill them. I suppose one shouldn't be proud of being a paid baby-killer, but in the case of mosquitoes I am rather proud of having served my country. In fact, some of my students and I have discovered a whole new way of attacking the little flying syringes: photodynamic dyes.

There are a lot of common dyes that exhibit something called the photodynamic effect. That is, they fluoresce under various light conditions. When some compounds are exposed to forms of electromagnetic radiation or light, the added energy kicks an electron loose. This loose electron quickly returns to its original position. Since it took energy to knock it loose, it releases energy when it falls back into place. This released energy is sometimes seen as light, and we call the phenomenon fluorescence.

However, this series of chemical changes sometimes causes, in these dyes, the release of oxygen. The oxygen isn't released as two oxygen atoms paired together, as usual, but as a single oxygen atom. These single oxygen atoms, also called oxidants, are very reactive and can attack other molecules in the vicinity and cause local de-stabilization. Oxidants can break apart cell membranes and perform other mischief-making activities. You and I also make these oxidants sometimes and that is why we need to have antioxidants in our diet. (Antioxidants combine with the oxidants and neutralize them so they don't cause damage.)

Scientists have taken advantage of this photodynamic fluorescence in many ways. These dyes can have a localized destructive effect on cancer cells if one injects the dye into the tumor, and then shines light where the dye has been absorbed. This has been used on various inoperable tumors in the brain where light usually isn't found. At least enlightenment is seldom found in my brain. Dr. Richard Heckmann, from Brigham Young University, and I were able to kill a type of parasite on fish gills using this technique.

So, some students at Mesa State College, now Colorado Mesa University, here at MSC have exposed the different stages in the life cycle of local mosquito, *Culex tarsalis,* to a dye called Rose Bengal. We did this in a dark box and then later exposed the mosquitoes to light. We found that the dye couldn't penetrate the eggs, so it had no effect. The larvae however, ingested the dye in the water as they fed and when exposed to light died very quickly. The pupae stage of mosquitoes does not feed. They are like the cocoon stage of other insects. So, if the pupae weren't exposed before they pupated, they were not killed. But we could feed adult mosquito's sugar water laced with Rose Bengal. When we did that and then exposed them to light, they also succumbed.

Now of course this is entirely useless information since we couldn't possibly put Rose Bengal in the water supply since it could affect all living things detrimentally. Maybe these dyes could be used in closed aquaria or something to treat fish diseases. But I have considered how entertaining it might be to drink some Rose Bengal, go out at night and let some mosquitoes take a blood meal, then shine a flashlight on them and watch them explode. Might give my cheeks a healthy, rosy glow as well. Of course, I'd have to stay out of the light myself until my body had metabolized the dye.

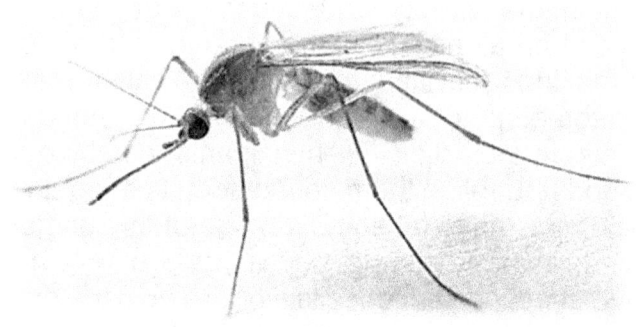

SKEETER MOUNTAIN

*There is a quiet that can be heard just before dark
In the mountains when the air is calm,
And light is growing dim
Often, at this time, campfires emit their first sparks
It's almost as if nature worships
With a tone of a natural hymn*

*Though the sound is faint like an ancient chant
It swells louder in the ear and
Then fades away to nothing
At first it brings peace and contentment, a pleasant descant
Only filled with unease and
Vague feelings of ancient stings*

*And as the darkness grows so does the fires glow
Just as fire burns
And rain turns to mud
There is something that lives in between to and fro
The hidden seeker
Of your blood*

Research Ideas

    The problem with flies and mosquitoes is that everyone thinks they are just one thing. The truth is there are numerous different species in any geographic area and most people don't know the difference. Yet almost every species has attributes, behaviors, and responses different from other species.

For example. Female mosquitoes that have mated often begin flying in a zig zag pattern at a specific time of day, seeking various attractants that will lead them to a blood meal. However, not all species search in the same way. Some cruise close to the ground, some fly higher up. Unmated females may not be out at all at the same time as each species, and each gender, may be seeking different food sources. Ferreting out these species' differences can be very interesting, challenging, and useful in understanding ecology, biology, and control efforts.

With so many types of flies and mosquitoes available it is easy to obtain specimens for research. Just leave a piece of liverwurst out on a windowsill for a little while and you can take your pick of flies. Mosquitoes can be easily captured with various simple carbon dioxide traps.

Because of their habits a lot of research has been done on food selection, life cycles, and disease transmission from unclean environments to humans. Some flies are very easy to raise in captivity and doing food preference studies and bionomic reactions to temperature, light, UV light, humidity and so forth are always welcome, especially on species that are less common. Following is a list of questions that could be asked and answered for nearly every species of fly. While common and disease related species may have much information, less common and benign species are simply not understood. There are many more.

1. What time of day does your dipteran seek food?
2. Does it have a favored direction?
3. Does it recognize color? Attraction or repelling?
4. Does it avoid certain odors or environmental cues?
5. What does it eat?
6. Do the two genders do the same things?
7. Where do they rest?
8. How far will the go?

9. Where do they mate?
10. Where do they breed?
11. How does temperature effect the various life cycle stages?
12. How does light effect various species, genders, and life cycle stages?
13. What types of blood cells are found in your species and in what proportions?'
14. Are the blood cells consistent across all stages?
15. What do their blood cells do?

CHAPTER 13 - MANTIDAE

Phylum Arthropoda
Class Insecta:
Order Mantodea:
Family Mantidae - 2,400 species
Species - Mantis religiosa

"Give us this day our daily bread . . ."

<u>*Milieu interieur*</u>

I bet very few people today know that rabbits, which are generally herbivores, produce cloudy urine that is basic in pH. I suppose my granddaughter knows this as she raises rabbits. But aside from a few rabbit fanciers and a couple of veterinarians, who knew? I realize one might well ask, "Who cares?"

I suppose you also are not sure why I am talking about rabbits to begin a chapter on a well-known insect called the praying mantis. It's because the Manidae are carnivores and rabbits led to a unique discovery about carnivores that might provide some ideas for research topics on praying mantises.

The truth is I didn't know rabbit urine is supposed to be cloudy and basic either until I read about Claude Bernard who did know this. Claude was a French physiologist in the first half of the 20th century. He is the one who knew this little tidbit about rabbit urine. So, he was intrigued when some rabbits, brought into the lab from a local market, proceeded to urinate on his table with a clear and acidic urine.

He also happened to know that most carnivores have clear and acidic urine, so he wondered if the rabbits had been eating hamburgers or something. He thought it more likely that they had not been fed recently and were in an advanced state of fasting. It dawned on him that a fasting animal might be digesting its own tissues, thus causing it to metabolize as if it were a carnivore. First, by feeding the rabbits a normal diet he observed a change to cloudy and basic urine. Then feeding the rabbits a meat diet, by trickery, he once again observed clear acidic urine, verifying his theory.

Somehow this discovery led him to propose the idea of the "milieu intérieur", now known as homeostasis. This is the idea that variables in the body are regulated so that internal conditions remain stable and relatively constant in the healthy individual. So basically, Claude is the guy who started making people pee in a cup when they go to the doctor. The doctor needs to know if he is dealing with an herbivore or a carnivore, I guess. I know I would treat the two kinds of patients differently.

> **I bet very few people today know that rabbits, which are generally herbivores, produce a cloudy urine that is basic in pH.**

It's commonly said that you are what you eat. But that just doesn't make sense. Do you know how many peanut butter and honey sandwiches I have eaten in my life?

And what about the praying mantis. While they commonly prey on all kinds of insects, they are not particular. They have commonly been observed eating hummingbirds, warblers, sunbirds, honeyeaters, flycatchers, and European robins. They have even been observed eating frogs and lizards. When they capture birds, they go straight for the brains and many species will decapitate and consumer their paramour after mating.

Praying mantises are ambush predators. They patiently stalk their prey and ambush them. When they strike, they do so with lightning speed, attacking with those big front legs so quickly that it's hard to see with the naked eye. In addition, they have spikes on their legs to skewer and pin the victims into place.

Mantises have long necks, upright posture, distinct faces, direct gaze, and are decidedly charismatic. But They are also masters of disguise. They often appear as leaves, sticks, or branches. But some mantises even molt at the end of a dry season to become black, conveniently aligning themselves with the brush fires that leave a blackened landscape. The flower mantises are crazy; some wildly ornate, others looking so convincing that unsuspecting insects come to collect nectar from them ... and become dinner in the meantime.

They can see in all directions, 360 degrees, have 3-d vision and an area of central focus so they can track prey precisely. They can sense bat echolocation signals. Bats will attack them, but they fight back ferociously and sometimes escape. They can jump with extreme precision and contort their body in midair in a very athletic manner. The ancient Greeks, Egyptians, and Assyrians all thought they had supernatural powers. Hence the genus name "mantis", which is a Greek word for prophet.

> From whence arrived the praying mantis?
> From outer space, or lost Atlantis?
> glimpse the grin, green metal mug at masks
> the pseudo-saintly bug, Orthopterous
> also carnivorous,
> And faintly whisper, Lord deliver us.
> - Ogden Nash -

But that is just biology. It turns out that Mantises have cultural and symbolic meaning far beyond their biological existence. They have even sparked a philosophical method of living called the "Mantis Way."

The Mantis Way

The mantis has been revered in numerous cultures, almost always for the same characteristics of peace, quiet, and calm. This strikes biologists who know that they are voracious carnivores, sometimes devouring their own mates as weird. But culture has never been particularly accurate or scientific.

Usually, the mantis seems to be invoked when people lives have been overcome with business, activity, and chaos. Presumably the Mantis can hear the still small voice within us that we cannot because of the noise we have generated. I think that's what it said in the article I read although there weren't any equations or graphs to clarify their meaning.

After observing this creature for any length of time people seem to see symbolism in the praying mantis stillness and patience. The mantis takes its time and seems to live at its own pace.

SYMBOLISM OF PRAYING MANTIS

- Stillness
- Awareness
- Creativity
- Patience
- Mindful
- Calm
- Balance
- Intuition

These traits have led the mantis to be a symbol of meditation and contemplation. In fact, in China, the mantis has long been honored for her mindful movements.

The mantis seems to never make a move unless sure it is the right move. This is a message many societies have chosen to emulate. It does seem like a good idea to contemplate and be sure our minds and souls all agree together about the choices we are making in our lives. Personally, I am alright with this idea although some humans debate the idea of a soul. Many of those doubting believe in an id. I don't know if mantises have souls or ids.

An appearance from the mantis is a message to be still, go within, meditate, get quiet, and reach a place of calm. It may also be a sign for you to be more mindful of the choices you are making and confirm that these choices are congruent. Of course, it may also represent a stalking tiger just before it leaps to destroy its prey.

The Kalahari Bushmen in Africa worship and consider the Praying Mantis as the oldest symbol of God. They believed it to be an incarnation of God, and whenever they would sight one, they would try and decipher its message. I don't know, maybe the Bushmen of Africa know something I don't.

Actually, I bet the bushmen of Africa know a lot of things that I don't. I probably wouldn't last two days in the bush, in spite of my advanced degrees in the biological sciences. Like my daughter once described it to her friend, "he's not the kind of doctor than can help anyone.

But as you can see, there are a lot of interesting things about the Mantidae.

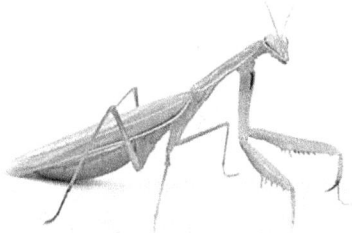

MANTIS RELIGIOSA

Serine and soulful in sincere devotion
Docile, slow, decidedly pure
Arms folded in supine affection
Every thought and action spiritually sure

Still contemplation as it worships alone
Slender aesthetic distinctive face
And direct gaze of its own
A symbol of both faith and grace

But they say that looks can deceive
A predator and a carnivore
That leaves others to grieve
Vicious voracious purveyor of gore

Research ideas for Mantises

Praying mantises are often not found in large numbers as they are carnivores, and there are never as many carnivores as there are prey animals. Because they will only eat live food that they have hunted and procured they can be difficult to raise in captivity. This makes them less available for the amateur to study.

However, there is still much that could be done seasonally by:
1. What are their diets?
2. Examine them for parasites; both ectoparasites and endoparasites.
3. Examining their blood parameters over a growing season.
4. Identifying the species in your area: including numbers, kinds, where they are found and what time of year.
5. Determine the average number of eggs in their ootheca.
6. Determine the average size and/or weight of each egg and ootheca.
7. The number of flies consumed by individuals could be determined for different species and areas.
8. Is cannibalism observed?
9. What is their natural diet?
10. Determine the time and conditions for hatching as well as the number of successful offspring occur.

CHAPTER 14 - ISOPODA

Phylum Arthropoda
Subphylum Crustacea:
Class Malacostraca:
Order Isopoda:
Family Armadillidiidae:
Genus Armadillidium - pill bugs

"Rolly Polly
Rolly Polly
The old diary
Long forgotten
Brush off the dust and remember
Your grin just as before
Your picture remains here
Then memories appear
In my heart"
Old Gaelic children's rhyme translated into English

<u>Bonus chapter</u>
  Well, this is a bonus chapter. See, the Isopoda are not insects at all, although I guess they would be considered creeping things. I am including them here because they are common in human environments, they are easy to keep, they are fascinating creatures, they play significant roles in the environment, and there is still much that is not known about them because they are not considered of economic importance.
  Whatever you call them, pill bug, roly-poly, woodlouse, armadillo bug, or potato bug, they are fascinating creatures. Most people associate them with dark, damp environments such as under rotting logs or rocks. However, they are nocturnal and present a very different picture at two in the morning. They are actually "good bugs" because they feed on rotting vegetation and return nutrients to the soil. They don't bite, sting, or attack living plants. And recently they have been shown to concentrate and extract heavy metal contaminants from soil.

Though we commonly refer to them as bugs, they are not insects but are crustaceans that breath through gills. That is why they require a moist environment for survival. They never urinate and can drink fluid from either their mouth or anal region. The young are carried in the mothers' pouch and so it sometimes appears that she is giving live birth. Their blood is blue, and they can eat metals in the soil, thus decontaminating the soil. I'm telling you; strange things happen in the night.

Categories

There are 5000 species of terrestrial Isopods. I didn't count them myself. I take the word of Isopodologists who, I presume, have. Isopods are roly-poly pill bugs and I have become interested in them for a variety of reasons. They are sort of land dwelling shrimp or lobsters.

Crustaceans are a completely different category than insects and are nearly as interesting as insects. But while scientists can't create insects, or crustaceans, we are pretty good at creating categories. Categorizing is something which scientists do well. Scientists are all about creating categories of similar, but different, things. Categorization has been quite useful in some respects, although I can't think of what they are right off the top of my head.

This has gotten me to thinking about categories of people. There aren't any, of course. That would be politically incorrect. Everyone is just like everyone else regardless of any outward characteristics they might possess such as gender or family relationships.

Some would say there is no difference between my mother-in-law and any other woman I might meet or associate with. All women should be viewed equally, and neither should be treated any differently. I believe this is an unscientific, and generally unwise, assumption.

Categories are created as soon as someone names something. I suppose categorization is a part and parcel with language. Ass soon as you name something, you can then name something else according to how similar or different the two appear to be. This is what scientists do all the time. They name something a Lepidopteran and soon discover that there are various types of Lepidopterans that seem similar but different. So, they create another category.

This works fine with things. It doesn't work so well with people. Naming people not only designates an object but also identifies a unique individual with characteristics that no one else shares. This new thing with a name has its own variability, freedom, prejudices, and tendencies. One can study a thing. One has a relationship with a person.

As a scientist, I have created three classes of people, but these classifications sometimes overlap. First, there are simply 'people'. This is the largest and most valuable class. These are the ones who make things, grow food, and provide electricity and other services.

'People' make certain assumptions that most 'people' share. For example, people might say that children are cute, twilight is pleasant and sentimental, and the underdog in a fight is always noble. Often 'people' cannot express these and other complex thoughts very eloquently. But when someone says, sadly, that it is a beautiful evening, we know what they mean. This is why there is a second category of people.

There is a smaller group of people that 'people' call 'poets'. 'Poets' generally share the same commonplace ideas and sentiments as 'people' but can express them in strange and delicate ways that capture what 'people' really mean when they say, "It is a beautiful evening."

Where I, a scientist, might comment, "The sun is going down", Henry Wadsworth Longfellow said ". . . And as the evening twilight fades away the sky is filled with stars, invisible by day." Where I might say, "Children are cute", Muhammad Ali said, "Children make you want to start life over."

'Poets' can express sentiments of the 'people' in an elevated and meaningful way that captures the complex, and often contradictory feelings in commonplace ideas. But they are still the sentiments of the 'people'. In general, 'Poets' differ from 'people' through their sensitivity and fluency. Also, their irresponsible behavior and poverty.

'Scientists', and other technical academicians, make up the third category. They are the ones who tell you that, "In reality, childhood is ugly and difficult." They also say that "The underdog is the helpless creation of injustice." 'Scientists' tell you that, "We are all just "animals" and that evenings are about the earth's rotation around the sun." 'Scientists' tend to differ from 'people' by their insensitivity.

The problem with categories, of course, is that nothing in human experience is all one thing. A mother-in-law is like the twilight: half one thing and half another. The underdog is a victim, but not helpless. Children are both charming and ruthless. And, of course, not all scientists are clumsy, insensitive oafs. In fact, not all poets write poetry.

There are scientists who speak as poets. This explains part of the popularity of Einstein who said, "Pure mathematics is, in its way, the poetry of logical ideas." In contrast, Stephen Hawking is brilliant, but no poet.

Like all categories, mine are incomplete. For example, I don't know what category to put people in who now say there are no categories. Then there are those who maintain that there are clear distinctions between people of different skin tones because some of them are helpless underdogs.

But even the categories of Isopods are constantly being revised. Categorization is an endless task. Become a scientist. There's great job security.

> There are 5000 species of terrestrial Isopods. I didn't count them myself. I take the word of Isopodologists who, I presume, have. They aren't bugs at all, but crustaceans. They are sort of land dwelling shrimp or lobsters.

Crustaceans
Much of the research on invertebrates as fertilizer, including the work on chitin's effects on plants has been done using marine crustacean such as crab, shrimp, and lobsters which are sometimes abundant in the oceans, and which make up an important human food element. Waste product from processing these creatures for food have had some attention as fertilizer for crops.

But one organism that is known as a relatively benign decomposer, has a chitinous exoskeleton, is easily reared on waste material, and is more closely related to marine organisms is the terrestrial isopod of the Family Armadillidium and Porcellio, the roly-poly bugs and sow bugs.

Because they are not usually crop pests, are intermediate hosts for only a limited number of minor parasites, and do not reproduce as rapidly as some insects, they have been somewhat neglected in research. But being terrestrial crustaceans, their chitin is more closely akin to the better known marine and aquatic crustaceans and because they can be grown on decaying organic matter such as waste-food, manure, or plant residual they may have utility as a source of fertilizer while fulfilling other important roles as decomposers and a sustainable agriculture program.

> **Because they are not usually crop pests, are intermediate hosts for only a limited number of minor parasites, and do not reproduce as rapidly as some insects, they have been somewhat neglected in research. But being terrestrial crustaceans, their chitin is more closely akin to the better known marine and aquatic crustaceans and because they can be grown on decaying organic matter such as waste-food, manure, or plant residual they may have utility as a source of fertilizer while fulfilling other important roles as decomposers and a sustainable agriculture program**

One additional difference between isopods and insects is that because of the species relationship, the isopod exoskeleton is far richer in calcium than most insect chitin.

Terrestrial Isopods have several other unique features that make them interesting.

- Like their marine cousins, terrestrial pill bugs use gill-like structures to exchange gases. They require moist environments to breathe but cannot survive being

submerged in water. Their gills are located on their ventral side behind their legs.
- Like all arthropods, pill bugs grow by molting a hard exoskeleton. But pill bugs don't shed their cuticle all at once. First, the back half of its exoskeleton splits away and slides off. A few days later, the pill bug sheds the front section. If you find a pill bug that's gray or brown on one end, and pink on the other, it's in the middle of molting.
- Like crabs and other crustaceans, pill bugs tote their eggs around with them. Overlapping thoracic plates form a special pouch, called a marsupium, on the pill bug's underside. Upon hatching, the tiny juvenile pill bugs remain in the pouch for several days before leaving to explore the world on their own.
- Most animals must convert their wastes, which are high in ammonia, into urea before it can be excreted from the body. But pill bugs have an amazing ability to tolerate ammonia gas, which they can pass directly through their exoskeleton, so there's no need for them to urinate.
- Though pill bugs do drink the old-fashioned way—with their mouthparts—they can also take in water through their rear ends. Special tube-shaped structures called uropods can wick water up when needed.
- Most kids have poked a pill bug to watch it roll up into a tight ball. In fact, many people call them roly-polies for just this reason. Their ability to curl up distinguishes the pill bug from another close relative, the sowbug. Interestingly, the tendency to curl up is stronger during the day when exposed to light. At night, they may just scurry away faster several times before curling up.
- Yes indeed, pill bugs munch on lots of feces, including their own. Each time a pill bug poops, it loses a little copper, an essential element it needs to live. To recycle this precious resource, the pill bug will consume its own poop, a practice known as coprophagy. However, this is also why they are good at removing heavy metals from contaminated soil.

- Like other animals, pill bugs can contract viral infections. If you find a pill bug that looks bright blue or purple, it's a sign of an iridovirus. Reflected light from the virus causes the cyan color. These viruses are not contagious or harmful to humans. In fact, they don't seem to have much deleterious effect on the isopods either.
- Many crustaceans, pill bugs included, have hemocyanin in their blood. Unlike hemoglobin, which contains iron, hemocyanin contains copper ions. When oxygenated, pill bug blood appears blue.
- Pill bugs are important for ridding the soil of heavy metal ions by taking in copper, zinc, lead, arsenic, and cadmium, which they crystallize in their midgut. Thus, they can survive in contaminated soil where other species can't.
- Pill bugs represent the only crustacean that has widely colonized the land. They're still a bit "fish out of water," though, as they are at risk of drying out on land; they haven't developed the waterproof waxy coating of arachnids or insects. Pill bugs can survive until they get down to 30 percent dry.
- If the humidity gets really high in the atmosphere, above 87 percent, pill bugs can absorb moisture from the air to stay hydrated or improve their hydration.
- Pill bugs probably came to North America with the lumber trade. European species may have originated in the Mediterranean region, which would explain why they don't survive winters where it gets below 20 degrees F as they aren't underground burrowers.

LATE NIGHT PILL BUGS

*Strange things happen in the night
When most folks are asleep in bed
If you venture forth when there is no light
And look beneath the rock behind the shed
You will find nothing there
Although you looked just this afternoon
And found pill bugs hiding from the glare
Tonight, they scamper beneath the moon
No flexing and rolling to play dead
But hurrying from here to there
There is much to do when all is said
And it must be done with the greatest care
Time is short until sunrise
When they must be home behind the shed
Without shelter they all will die
When folks awake, they must be in bed
Strange things happen in the night*

Research ideas for pill bugs

Terrestrial Isopods may be one of the easier organisms for research. They are found in many environments, even in urban settings. They are easy to maintain in captivity and probably play a significant role in soil biology. They are not dangerous to handle and in fact are rather fun to play with.

Because they lack medical, agricultural, or environmental significance their study is somewhat neglected.

However, they probably play a significant role in a number of ways in the soil environment. This may be of some importance in the urban environment as the areas where them may be found may be pockets of biological life that are unique. Multiyear studies may also show some effects of things such as climate change or local environmental conditions.

Some possible studies done with terrestrial Isopods of various kinds.
1. Does their presence have a beneficial effect on plants? What?
2. Can their exoskeletons benefit plants? Compare species and subtypes.
3. How do they effect the soil fauna where the reside?
4. What effect does temperature, substrate, moisture, gasses, and other environmental effects have on them?
5. What effect does isolation have?
6. How do they respond to density?
7. What do they do by the hour of the day?
8. What effect does pollution have?
9. Does the distance from air pollution, such as numbers at various distances from highways have?
10. Can they complete a maze?
11. Do they respond to light, vibration, color, magnetic fields, electrical fields?
12. Food preferences?
13. Parasites?
14. How does their presence alter the environment in matters such as temperature, moisture, gasses, or substrate?
15. Are their numbers and locations related to things such as highway exhaust, pollution, minerals and such.

CHAPTER 15 - CONCLUSIONS

Insects are not drawn to the flame.
They are drawn to the light.
They plunge into the flame to get to the light.
Then all that is left is moth smoke.
- Gary McCallister -

If you thought this book was about insects, you have partially missed the point. Sure, insects are emphasized and are amazing creatures. But this book is also about the magnificence of nature and creation.

Do you really believe the complexity and organization of these invisible systems evolved without some unimaginable intelligence or force. I know there are those that believe that, but I suspect they have other reasons for their beliefs than reality. So a major part of this book is to illustrate this interest and complexity of the creeping things. God called them good.

But the other purpose of this book is to celebrate you. Step outside. Draw a circle around you with a ten-foot radius. Inside this circle are four hundred things no one understands. In fact, some no one has ever even seen. Can you understand them? Can you find them?

Could our lives be made more interesting if we paid more attention to our world. Perhaps insects don't interest you. Do electronics? How about gardening? Did you know you could buy a telescope for stargazing, or a microscope for studying soil and water for about the same price as a set of good golf clubs. Protozoa? Nematodes, they're everywhere. And you wouldn't have to travel or pay green fees. You can do both with friends just as easily.

Like to party? How about International Moth Night in July. You could sit around all night in your backyard collecting and photographing moths as they visit your moth trap.

Stop letting the professional gate keepers of science from letting you have the same fun they get to have. You don't need their license. You don't need their permission. You never gave them yours. And you don't need their approval. Learn about your world, your way, when and where you want to or can. Take science back again.

There is full-time employment for anyone and everyone simply in exploring the world without destroying it. In doing so we may begin to understand something of its marvelous richness and complexity. I believe we'll also begin to see that it does have uses that we never suspected and that its main value is what comes to us directly from mere coexistence with living things.

I believe you will discover that the natural world has an impact on our minds and bodies, subtle and powerful, that goes far beyond the advantages of converting all things to cash and calories. And we may even begin to get a glimpse of Gods grandeur, majesty, and intelligence in the complexities and creativeness of the universe, and our earth in particular.

The beauty of nature around us can produce some of the most inspiring and delightful experiences in life. The discoveries you will find will create emotions and a deep sense of gratitude for our Heavenly Father for this magnificent world.

This earth is not an end in itself. It is part of God's plan for His children. The physical earth was necessary in order for our minds to create parables and stories that clarify and enhance meaning in our lives. Its purpose is to provide the setting in which men and women can live and experience agency, bondage, joy, sorrow, and learn what they need to progress both spiritually and physically.

This earth was created for our benefit, and it is living proof of the love of God. The Lord declared, "Yea, all things which come of the earth ... are made for the benefit and the use of man, both to please the eye and to gladden the heart."

These divine gifts do not come without duties and responsibilities. Throughout Christian scriptures these duties are repeatedly captured by making references to "stewardship" or "shepherding." These terms suggest a responsibility to take care of something that belongs to God for which we are accountable.

We often think of our lives as belonging to us. Yet we had no part in creating ourselves. Our previous parents created us, and God created our first parents. In addition, everything we need to live and prosper is already in place for our use as we learn how to utilize these raw materials. Some people say there is no free lunch but in truth everything that we need is already in place, for free as far as we are concerned.

We did not create the earth. We do not have even an inkling of the power it would take to do so. We have no claim. "I, the Lord, stretched out the heavens, and built the earth, my very handiwork; and all things therein are mine." All that is on the earth belongs to God, including our families, our physical bodies, and even our very lives.

We are caretakers, set to dress and take care of it. We are guardians of his divine creation. The Lord said that He made "every man accountable, as a steward over earthly blessings, which I have made and prepared for my creatures."

We are allowed to use earthly resources according to our own free will. This confuses many men who then think they have done something themselves. In truth they have but it is like the many who inherits a farm for free, then sells it to build an empire chain of grocery stores. He forgets that he started with something of great value that was gifted to him. He may forget that he has a responsibility to use his estate wisely.

"And it pleaseth God that he hath given all these things unto man; for unto this end were they made to be used, with judgment, not to excess, neither by extortion." And as President Russell Russell M. Nelson once remarked: "As beneficiaries of the divine Creation, what shall we do? We should care for the earth, be wise stewards over it, and preserve it for future generations."

What can we do to be good stewards or shepherds? Considering our individual circumstances, we can:
- use resources reverently and prudently.
- be supportive of community efforts to protect the earth.

- Adopt conservative lifestyles that respect life.
- make our own living spaces tidier, more beautiful, and more inspirational.

And we can learn everything we can about the earth and how it operates, right down to the creeping things that are the foundation of all life.

We have the opportunity to participate in His creative work. We can cultivate the earth, add out own constructions, and show respect to God's creations no matter how small or creepy. We contribute to creation through art, architecture, music, literature, culture, and those things that beautify and embellish our world.

We also contribute through scientific and medical discoveries that preserve the earth and life upon it. President Thomas S. Monson summarized this concept with these beautiful words: "God left the world unfinished for man to work his skill upon ... that man might know the joys and glories of creation." When we take what God has created and use it to create our own world we are participating in re-creation, although we usually spell and pronounce it <u>recreation.</u>

Our role as stewards of earthly creations is not solely about conserving or preserving them. The Lord expects us to work diligently, as moved upon by His Holy Spirit, to grow, enhance, and improve upon the resources He has entrusted to us—not for our benefit only but to bless others. Great spiritual blessings are promised to those who love and care for the earth and their fellow men and women. And great satisfaction comes from learning how good the earth is, even the creeping things.

ABOUT THE AUTHOR

Dr. Gary L. McCallister is a practicing biologist with four children and eighteen grandchildren. He practiced for over forty years and finally gave up and retired. He has been awarded a participation trophy called Emeritus standing by the Colorado Mesa University.

He is a highly trained professional windbag and can prove it with sixty-two insignificant scientific papers, seventeen books, and an award-winning weekly science column for the newspaper. Okay, it was for scientific humor, a category heretofore unrecognized. He has also produced twenty-one music CDs of uneven quality and is a luthier of the mountain dulcimer. His unique designs are popular throughout the western United States where he has been forced to give most of them away.

He lives in southwest Idaho where he tends bees, grows prickly pear cactus, and plays music on his guitars, mandolins, banjos, and mountain dulcimers. Yet, he still dabbles in the biology of creeping things. You can contact him at: mccallistergaryl@gmail.com

www.ingramcontent.com/pod-product-compliance
Lightning Source LLC
Chambersburg PA
CBHW052316220526
45472CB00001B/147